Lecture Notes in Computer Science 15945

Founding Editors

Gerhard Goos

Juris Hartmanis

The series Lecture Notes in Computer Science (LNCS), including its subseries Lecture Notes in Artificial Intelligence (LNAI) and Lecture Notes in Bioinformatics (LNBI), has established itself as a medium for the publication of new developments in computer science and information technology research, teaching, and education.

LNCS enjoys close cooperation with the computer science R & D community, the series counts many renowned academics among its volume editors and paper authors, and collaborates with prestigious societies. Its mission is to serve this international community by providing an invaluable service, mainly focused on the publication of conference and workshop proceedings and postproceedings. LNCS commenced publication in 1973.

Gidon Ernst · Kristin Yvonne Rozier

Editors

Model Checking Software

31st International Symposium, SPIN 2025
Hamilton, ON, Canada, May 7–8, 2025
Proceedings

 Springer

Editors
Gidon Ernst [iD]
Ludwig-Maximilians-Universität München
Munich, Germany

Kristin Yvonne Rozier [iD]
Iowa State University
Ames, IA, USA

ISSN 0302-9743   ISSN 1611-3349 (electronic)
Lecture Notes in Computer Science
ISBN 978-3-032-06846-0   ISBN 978-3-032-06847-7 (eBook)
https://doi.org/10.1007/978-3-032-06847-7

Preface

This volume contains the papers presented at SPIN 2025: 31st International Symposium on Model Checking Software (SPIN), held on May 7–8, 2025 in Hamilton, Canada. SPIN 2025 was co-located with the ETAPS International Joint Conferences on Theory and Practice of Software, hosted by McMaster University. This edition of SPIN was the 31st instalment of the symposium.

The symposium sought submissions on topics related to formal verification for automated analysis of (concurrent) software/hardware, including model checking, deductive verification, automated theorem proving (including SAT and SMT solving), abstraction and symbolic execution techniques, static analysis and abstract interpretation, modular and compositional verification techniques, verification of timed and probabilistic systems, automated testing using advanced analysis techniques, program synthesis, derivation of specifications and test cases via formal analysis, formal specification languages, temporal logic, design-by-contract, formal analysis of learned systems, or any combination of these. Papers could be submitted either as regular papers, as full tool papers, or as short papers.

SPIN 2025 attracted 20 submissions, of which there were 15 research papers, 4 full tool papers, and one short paper. Each submission was reviewed by three program committee members in a single-blind manner. After discussions, 9 papers were accepted, including one full tool paper. The contribution "0–1 Laws for LTL and CTL over Random Transition Systems" by Yanni Dong, Milan Lopuhaä-Zwakenberg, and Marielle Stoelinga received the SPIN 2025 best paper award.

For the second year, SPIN featured an artifact evaluation track to foster tool development and reproducibility. This track was chaired by Julie Cailler and Nian-Ze Lee. The artifact evaluation committee was responsible for reviewing artifacts submitted by authors to support the contributions in their papers according to the EAPLS guidelines. Artifact submission was mandatory for full tool papers alongside paper submission and optional at a later deadline for all other papers. SPIN 2025 received 7 artifact submissions, of which 4 correspond to accepted papers, that were awarded one or more badges for being available, reusable, and/or functional.

The program included two fantastic keynote talks: Orna Grumberg gave an account of "Developments in Symbolic Model Checking" and Alexandre Duret-Lutz talked about Spot—"Growing an ω-automata library." Both keynotes were very insightful, entertaining, and well-attended. SPIN featured a special session on MoXI as a common interchange format for model checking tasks, with presentations by Kristin Yvonne Rozier, Cesare Tinelli, Chris Johannsen, Rob Lorch, Po-Chun Chien, and Nicola Gigante, leading to stimulating discussions. Moreover, the results of the annual competition on Constrained Horn-clause solving (CHC-COMP 2025) were presented. We are grateful to all speakers at the symposium for sharing the findings of their research and their views on the field.

Finally, we would like to thank several organizations and people. First and foremost, we extend our heartfelt thanks to all PC members and external reviewers for their critical and timely reviewing, all the authors for their submitted work, and all attendees for participating. In addition, we thank the artifact evaluation chairs and committee, and we are very grateful to the local organizers of ETAPS 2025 and the student volunteers. We are also grateful to EAPLS for use of their artifact badging scheme, to EasyChair for hosting the review platform, and to Springer for publishing these proceedings.

May 2025

Gidon Ernst
Kristin Yvonne Rozier

Organization

Program Chairs

Kristin Yvonne Rozier	Iowa State University, Ames, USA
Gidon Ernst	Ludwig-Maximilians-Universität München, Germany

Artifact Evaluation Chairs

Julie Cailler	University of Lorraine & Inria, France
Nian-Ze Lee	LMU Munich, Germany & National Taiwan University, Taiwan

Steering Committee

Anton Wijs	Eindhoven University of Technology, The Netherlands
Stefan Leue	University of Konstanz, Germany
Susanne Graf	Verimag, France
Gerard Holzmann	Nimble Research, USA
Jaco van de Pol	Aarhus University, Denmark
Neha Rungta	AWS, USA
Willem Visser	Stellenbosch University, South Africa
Georgiana Caltais	University of Twente, The Netherlands
Christian Schilling	Aalborg University, Denmark

Program Committee

Marek Chalupa	Institute of Science and Technology, Austria
Marie-Christine Jakobs	Ludwig-Maximilians-Universität München, Germany
Nancy Day	University of Waterloo, Canada
Rohit Dureja	IBM Corporation, USA
Marie Farrell	University of Manchester, UK

Susanne Graf	Université Joseph Fourier, CNRS, Verimag, France
Ahmed Irfan	SRI International, USA
Violet Ka I Pun	Western Norway University of Applied Sciences, Norway
Dominik Klumpp	Albert-Ludwigs-Universität Freiburg, Germany
Ondřej Lengál	Brno University of Technology, Czech Republic
Cong Liu	Collins Aerospace, USA
Jianwen Li	East China Normal University, China
Lucas Martinelli Tabajara	Amazon, USA
Radu Mateescu	Inria Grenoble - Rhône-Alpes, France
Thomas Neele	Eindhoven University of Technology, The Netherlands
Marjan Sirjani	Mälardalen University, Sweden
Nestan Tsiskaridze	Standford University, USA
Masaki Waga	Kyoto University, Japan
Anton Wijs	Eindhoven University of Technology, The Netherlands
Matthias Volk	Eindhoven University of Technology, The Netherlands
Shufang Zhu	University of Liverpool, UK

Artifact Evaluation Committee

Ryan Dancy	University of Oxford, UK
Chiao Hsieh	Kyoto University, Japan
Matthias Kettl	Ludwig-Maximilians-Universität München, Germany
Levente Bajczi	Budapest University of Technology and Economics, Hungary
Thomas Lemberger	Ludwig-Maximilians-Universität München, Germany
Chen-Kai Lin	Institute of Information Science, Academia Sinica, Taiwan
Landon J. Taylor	Utah State University, USA
Mohit K. Tekriwal	Lawrence Livermore National Laboratory, USA
Aline Uwimbabazi	Schlumberger Foundation & Inria, France
Aosen Xiong	University of Waterloo, Canada
Zsófia Ádám	Budapest University of Technology and Economics, Hungary

Additional Reviewers

Milan Ceska
Crystal Chang Din
Masoud Ebrahimi
Zahra Moezkarimi
Áron Ricardo Perez-Lopez
Wendelin Serwe
Khashayar Etemadi Someoliayi

Developments in Symbolic Model Checking

Orna Grumberg

Computer Science Department, Technion, Israel

Abstract. Model checking is a powerful verification technique that automatically determines whether a given software or hardware system satisfies its specification. While it has achieved significant success, ongoing research aims to extend its capabilities to handle more complex systems and properties. One promising direction is symbolic model checking, which reformulates the verification problem into a logical form, leveraging the progress made in satisfiability solvers. Initially, model checking for hardware was encoded using propositional logic and solved using SAT solvers. Later, software model checking evolved to use first-order logic, analyzed with SMT solvers. More recently, Constrained Horn Clauses (CHCs) have proven effective for verifying program correctness with respect to safety properties. In this talk, we will introduce CHCs and explore their use in several unconventional applications that go beyond traditional safety verification.

Growing an ω-Automaton Library

Alexandre Duret-Lutz

LRE, EPITA, Le Kremlin-Bicêtre, France
adl@lrde.epita.fr

Abstract. This is an **invited talk** about the past, present, and future of Spot, a C++ library for the manipulation of linear-time temporal logic and ω-automata, with applications to model checking and reactive synthesis. Today, most of the features of Spot revolve around its support for transition-based Emerson-Lei automata. I discuss key developments in the history of Spot, and recently introduced features that I find exciting.

1 Present: What is Spot?

Today, Spot is presented as a C++17 library for the manipulation of linear-time temporal logic (LTL) and ω-automata. Calling Spot a C++ library is a bit reductive, because users can actually use three interfaces: the C++ library, the Python bindings, and a set of command-line tools. The Python bindings are typically used for interactive exploration, prototyping, and scripting. C++ is for production code, where efficiency matters. The command-line tools are designed to follow the Unix philosophy of using text streams as input/output to enable composition of tools using pipes. These tools may be used for scripting tasks, or help with benchmarking algorithms. Spot also offers certain specialized tools (like ltlcross and autcross that help to test LTL or automata-based algorithms.

2 A Brief History in Four Steps

2.1 Infancy

Spot started as a Master's and then PhD project. In 1999, Jean-Michel Couvreur had published two algorithms related to the automata approach for LTL model checking [5]: a translation from LTL to Transition-based Generalized Büchi Automata (TGBA), and an emptiness check for TGBA. Model checking is traditionally done using Büchi automata (BA), where accepting runs have to visit infinitely often a state that belongs to some acceptance set. TGBAs generalize those automata by using accepting transitions instead of accepting states, and by using multiple acceptance sets. As TGBAs can be more compact than BAs, we wanted to explore their use for model checking.

The automata-theoretic approach to model checking can be described in four steps: (1) translate some LTL specification φ into an automaton $\mathcal{A}_{\neg\varphi}$ that recognizes any behavior not allowed by φ, (2) build a Kripke structure S representing the state-space of the model that you want to check against φ, (3) build the product $\mathcal{A}_{\neg\varphi} \times S$ representing all behaviors of S that are disallowed by φ, and finally (4) ensure that this product is empty.

Spot used the definition of TGBA as an abstract interface between all steps, in a way that allowed on-the-fly computations. Steps (1) and (4) correspond to Couvreur's algorithms, and step (3) is easy to implement. That leaves us with step (2), which depends on the language used to express the model. Since some colleagues were working on Petri-Net tools, we built a couple of interfaces to expose the reachabilty graphs of Petri Nets constructed by those tools using Spot's TGBA interface.

Spot was built as a library to ease the combination of such user-supplied implementation of step (2) with the rest of the steps it provides [8].

Back them, Spot did not offer any command-line tools to its users, but it had a couple of such tools in its test suite. For instance one tool was called ltl2tgba because it was meant to test step (1). However, as a testing tool, it grew a *"junk drawer"* interface: a confusing set of cryptic options added without any logic, for the purpose of testing new features as they were implemented.

Of course, people started using this tool nonetheless, because it had some options to translate LTL into Büchi automata and print them in useful formats. But the set of options made it very difficult to use for people who knew the kind of output they wanted but not how to obtain it.

2.2 Spot 1.0: Introducing Command-Line Tools for Users

Seeing people benchmarking ltl2tgba (the testing tool) without always using the right options was a strong motivation to implement some command-line tools with a *sane* user-interface. The old ltl2tgba tool was eventually renamed ikwiad (" *I Know What I Am Doing*") to make sure people would switch to the brand new ltl2tgba. This one has a more friendly interface where you specify the type of output you want, and let the tool figure out what algorithms to chain to obtain that. Other useful tools were provided along the way, like genltl (for generating lists of LTL formulas used in various benchmarks), ltlfilt (for filtering streams for formulas, or ltlcross (for testing LTL translators) [6].

In order to interface with other tools, Spot already had parsers for several automaton formats. For some work on SAT-based minimization of Deterministic TBA [1] we had to write *yet another* parser to read Rabin automata produced by ltl2dstar [10]. Meanwhile, other teams were working on building automata with different acceptance conditions, and were inventing their own format, making it difficult to combine or compare tools. At ATVA'13 in Hanoi, we gathered with a few tool authors and started to discuss the need for some common and flexible format. We contacted more people after the conference to draft and implement what would eventually be called the *Hanoi Omega Automaton (HOA) Format* [2].

2.3 Spot 2.0: Building support for Emerson-Lei acceptance

The HOA format represents ω-automata in which the acceptance conditions are arbitrary Boolean formulas over atoms like Fin(⓪) (⓪ should be seen finitely often) or Inf(⑤) (⑤ should be seen infinitely often). Such acceptance conditions, also known as Emerson-Lei conditions, can represent all traditional ω-automata acceptances, but give freedom to introduce new acceptance conditions when needed. For example, the product of two automatas with acceptance conditions α_1 and α_2 assuming, w.l.o.g., disjoint colors) can be built with a simple synchronous product, setting the resulting acceptance to $\alpha_1 \wedge \alpha_2$.

Supporting HOA required a complete redesign of Spot, yielding version 2 [7]. Leaving the world of TGBAs, Spot started to provide algorithms that output more complex acceptance conditions (such as Safra-based determinization). It also opened many research objectives, to build algorithms that support generic acceptance conditions. Notable algorithms are our generic emptiness check [3], and our implementation of the Alternating Cycle Decomposition [4].

A second noteworthy feature of Spot 2.0 was its support for Jupyter's rich display system. Combined with Spot Python's bindings, it allows calling Spot's algorithms in Jupyter notebooks and getting some visual result. Such notebooks are great for prototyping algorithms, interactive exploration, teaching...

2.4 A Shift Towards Reactive Synthesis

With the support for HOA, Spot has grown into an ω-automata library, where model checking is just one *possible* application. In 2018, Maximilien Colange introduced a new application for Spot: LTL Reactive Synthesis. A new tool was added for this purpose ltlsynt [11]), and its development had ripple effects: several algorithms (e.g., our Safra-based determinization) were seriously improved, but also new concepts such as games, Mealy machines, and And Inverter Graphs (AIG) were eventually introduced in Spot [9].

3 Upcoming features

Recently, there have been a lot of developments to apply Reactive Synthesis to LTL f (LTL over finite traces). Doing this efficiently required the introduction of DFAs represented using Multi-Terminal Binary Decision Diagrams, similar to the DFA representation of Mona. We plan to try to generalize such a representation to automata with Emerson-Lei acceptance in the future.

References

1. Baarir, S., Duret-Lutz, A.: Mechanizing the Minimization of deterministic generalized Büchi Automata. In: Ábrahám, E., Palamidessi, C. (eds.) FORTE 2014LNCS, vol. 8461, pp. 266–283. Springer, Heidelberg (2014). https://doi.org/10.1007/978-3-662-43613-4_17

2. Babiak, T., et al.: The Hanoi Omega-automata format. In: Kroening, D., Păsăreanu, C. (eds.) CAV 2015. LNCS, vol. 9206, pp. 479–486. Springer, Cham (2015). https://doi.org/10.1007/978-3-319-21690-4_31

3. Baier, C., Blahoudek, F., Duret-Lutz, A., Klein, J., Müller, D., Strejček, J.: Generic emptiness check for fun and profit. In: Chen, Y.F., Cheng, C.H., Esparza, J. (eds.) ATVA 2019. LNCS, vol. 11781, pp. 445–461. Springer, Cham (2019). https://doi.org/10.1007/978-3-030-31784-3_26

4. Casares, A., Duret-Lutz, A., Meyer, K.J., Renkin, F., Sickert, S.: Practical applications of the alternating cycle decomposition. In: Fisman, D., Rosu, G. (eds.) TACAS 2022. LNCS, vol. 13244, pp. 99–117. Springer, Cham (2022). https://doi.org/10.1007/978-3-030-99527-0_6

5. Couvreur, J.M.: On-the-fly verification of linear temporal logic. In: Wing, J.M., Woodcock, J., Davies, J. (eds.) FM 1999. LNCS, vol. 1708, pp. 253–271. Springer, Heidelberg (1999). https://doi.org/10.1007/3-540-48119-2_16

6. Duret-Lutz, A.: Manipulating LTL formulas using Spot 1.0. In: Van Hung, D., Ogawa, M. (eds.) Automated Technology for Verification and Analysis. LNCS, vol. 8172, pp. 442–445. Springer, Cham (2013). https://doi.org/10.1007/978-3-319-02444-8_31

7. Duret-Lutz, A., Lewkowicz, A., Fauchille, A., Michaud, T., Renault, É., Xu, L.: Spot 2.0—A Framework for LTL and -Automata Manipulation. In: Artho, C., Legay, A., Peled, D. (eds.) ATVA 2016. LNCS, vol. 9938, pp. 122–129. Springer, Cham (2016). https://doi.org/10.1007/978-3-319-46520-3_8

8. Duret-Lutz, A., Poitrenaud, D.: Spot: an extensible model checking library using transition-based generalized Büchi automata. In: Proceedings of the 12th IEEE/ACM International Symposium on Modeling, Analysis, and Simulation of Computer and Telecommunication Systems (MASCOTS 2004), pp. 76–83. IEEE Computer Society, The Netherlands (2024). https://doi.org/10.1109/MASCOT.2004.1348184

9. Duret-Lutz, A., et al.: From Spot 2.0 to Spot 2.10: What's New? In: Shoham, S., Vizel, Y. (eds.) CAV 2022. LNCS, vol. 13372, pp. 174–187. Springer, Cham (2022). https://doi.org/10.1007/978-3-031-13188-2_9

10. Klein, J., Baier, C.: On-the-fly stuttering in the construction of deterministic ω-automata. In: Holub, J., Žďárek, J. (eds.) CIAA 2007. LNCS, vol. 4783, pp. 51–61. Springer, Heidelberg (2007). https://doi.org/10.1007/978-3-540-76336-9_7

11. Michaud, T., Colange, M.: Reactive synthesis from LTL specification with Spot. In: Proceedings of the 7th Workshop on Synthesis, SYNT@CAV 2018. Electronic Proceedings in Theoretical Computer Science (2018)

Contents

TRANSVER:
A Modular Program-Transformation Framework for Reduction to Reachability

Dirk Beyer , Marek Jankola , Marian Lingsch-Rosenfeld ,

Tian Xia , and Xiyue Zheng

LMU Munich, Munich, Germany

Abstract. Software verification is a complex problem, and verification tools need significant tuning to achieve high performance. Due to this, many verifiers choose to specialize on basic reachability properties. Instead of implementing algorithms for each possible specification, some verifiers implement known transformations from the given specification to reachability on their internal representations. Unfortunately, those internal transformations are not reusable by others. To improve this situation, we propose TRANSVER, a tool which offers transformations as modular stand-alone component, modifying the input program instead of the internal representation, enabling their usage as a preprocessing step by other verifiers. This way, we separate two concerns: improving the performance of reachability analyses and implementing efficient transformations of arbitrary specifications to reachability. We implement the transformations in a framework that is based on *instrumentation automata*, inspired by the BLAST query language. In our initial study, we support three important concrete specifications for C programs: *termination*, *no-overflow*, and *memory cleanup*. We conduct experiments with ten different verifiers. The experiments evaluate the efficiency and effectiveness of our transformations. The results are promising: Our transformations can extend existing verifiers to be effective on specifications for which they have no integrated support, and the efficiency is often similar or better to state-of-the-art verifiers that have integrated support for the considered specifications.

Keywords: Specification · Monitoring · Specification Reduction · Reachability · Formal Verification · Model Checking · Software Verification · Program Analysis

1 Introduction

Software verification is the problem to decide, for a given program P and specification φ, whether the program P satisfies its specification φ, in short: $P \models \varphi$. To address this issue, a software verifier is usually constructed in one of two ways: (a) implement a verification algorithm for $P \models \varphi$ or (b) *reduce* the problem by transformation in order to solve it using an existing algorithm. The transformation-based approach consists of two steps and assumes an existing verifier v that supports a specification φ'. In the first step, the problem $P \models \varphi$ is transformed to a problem $P' \models \varphi'$, such that $P \models \varphi$ holds if and only if $P' \models \varphi'$ holds. In the second step, the problem $P' \models \varphi'$ is solved by the existing verifier v.

© The Author(s) 2026
G. Ernst and K. Y. Rozier (Eds.): SPIN 2025, LNCS 15945, pp. 1–24, 2026.
https://doi.org/10.1007/978-3-032-06847-7_1

Several verification tools already choose to use transformations [1, 2, 3], since transformations allow for a separation of concerns: (1) support a rich set of specifications and (2) tune the performance of specialized algorithms for one particular specification. For example, given a verifier that supports reachability, the verifier can be extended to support other specifications, like termination and no-overflow, for which a transformation to reachability is available.

Currently, a developer who desires to extend a verifier to support more specifications using the transformation approach has to re-implement the transformation. The goal of this paper is to show (a) that it is possible to construct verifiers in a *modular* way from independent components, that is, compose an 'off-the-shelf' transformation with an 'off-the-shelf' verifier, such that a transformation can be used with arbitrary verifiers for C programs to support more specifications, (b) that transformation-based approaches can be even more efficient than integrated support for specifications, and (c) that the standalone transformations do not necessarily lead to a performance decrease in comparison to tool-specific implementations for checking the input specification (using internal transformation).

To achieve this, we developed TRANSVER, a transformation framework and tool for software verification, in which verification engineers write specifications at a high-level as *instrumentation automata (IA)*, which contain instructions to instrument and monitor the program. Instrumentation automata are inspired by the BLAST query language [4] and SLIC [5], which is a specification language for SLAM [6]. Both specification languages are based on monitor automata and have shown to be useful in practice, since the convenient and succinct notation of monitor automata is often easier to understand than LTL formulas. Monitor automata can be implemented either by instrumentation of the monitor into the program code [4, 5] or by an implicit on-the-fly product construction in which the monitor is a separate analysis component [7, 8]. Instrumentation automata observe the *control-flow automaton* of a program, and weave monitoring instructions and error assertions into the control-flow where appropriate.

To showcase our approach, we have implemented three transformations that reduce verification problems for which the specification is to check *termination*, *memory cleanup*, or (arithmetic) *no-overflow*, to verification problems for which the specification is to check reachability. Those three kinds of specifications are important and often used, because we want programs to not cycle infinitely but progress with useful computation (termination), we want the programs to free all allocated memory such that we can use the programs as components as part of other programs (memory cleanup), and we want the software to not run into undefined behavior, like signed-integer overflow, and always have values within their types (no-overflow). The target specification is reachability since most verifiers (28 verifiers) for software verification that participated in the competition on software verification (SV-COMP) in 2024 [9] support it and have focused on tuning their algorithms towards best performance on such tasks. Fewer verifiers support other specifications, such as *no-overflow* (19 verifiers), *termination* (16 verifiers), or *memory cleanup* (21 verifiers), since adding support for multiple properties is a time-consuming process.

In summary, the advantage of our approach is a clear separation of concerns, because the concern of optimizing a verification algorithm (for reachability) is now completely independent from checking whether a (non-reachability) specification is satisfied. In addition, this approach opens new opportunities, for example, using test-generation tools like fuzzers to check for violations of a specification defined as instrumentation automaton. For example, it is now possible to construct a fuzzing-based non-termination checker similar to existing tools [10, 11] without spending any development effort on the fuzzing tool. All it takes is developing an instrumentation automaton for TRANSVER. In particular, the verifiers EMERGEN-THETA [12], THETA [13], and THORN used TRANSVER as a preprocessing step for SV-COMP 2025, which demonstrates TRANSVER's modularity.

Contributions. This paper makes the following contributions: [1]

- We propose a verifier-independent, modular transformation framework for the instrumentation of C programs. It allows verifiers for reachability to be used also for other specifications as well.
- We provide an open-source implementation of our transformation framework.
- We conduct an experimental evaluation on the SV-COMP benchmark set of C programs, which shows that we can effectively extend existing verifiers to specifications that they did not support before (RQ 1), that verifiers combined with transformations sometimes even outperform verifiers specialized in verifying the original property (RQ 2, RQ 3), and that verifiers with internal transformations are not necessarily more efficient than a composition using our transformation framework and an 'off-the-shelf' verifier (RQ 4).

2   Related Work

Program transformations have a wide variety of applications [15, 16]. We focus on three kinds [17] of program transformations closely related to our approach.

Reducers can simplify complicated language constructs, for example, by sequentializing concurrent programs [18, 19], by reducing the program to a simplified syntax [1, 20], or by merging multiple loops into one single loop [21, 22]. There are also reducers that replace program constructs (for example, loops) by constructs that are easier to verify [23, 24, 25, 26, 27, 28]. Sometimes they even use information from run-time verification to ease the static analysis [29].

Specifications transformers convert a problem $P \models \varphi$ to a new problem $P' \models \varphi'$. This makes it possible to use algorithms for the verification of φ' to also verify φ [2, 30, 31, 32, 33, 34, 35, 36, 37]. Specifications transformers can also be used for testing, in order to transform a program and a coverage specification to another program and coverage specification, such that existing tools for test generation or test-suite analysis can be used [38, 39]. Our work focuses on this kind of program transformation. We improve over existing works by two aspects: modularity and generality. Outputting a modified C program makes the application of any C verifier that supports reachability effortless. Moreover, we demonstrate that our framework supports transformations for multiple properties.

---

[1] A preliminary version of this article was published as technical report [14].

Instrumentors add code to the program that is used to collect information for further analysis. Some instrumentors express the verification goal as part of the program to be verified. This has been studied for a variety of applications. One such example is in the context of verification witnesses [40, 41] in the case of MetaVal [42], which creates a product of the witness and the program. Another example is proof-carrying code [43], where the proof is embedded into the program. Furthermore, instrumenting additional logic into the program allows for its run-time monitoring [44, 45, 46] or improves the verification process, for example by using shadow memory [47, 48]. Furthermore, there also exist configurable instrumentors, for example, tools like AspectJ [49] and AspectC++ [50] are used in aspect-oriented programming. Moreover, there are verifiers with configurable instrumentation engines that can produce instrumented code. For example, SYMBIOTIC [51, 52, 53] can instrument LLVM and ESBMC [54, 55] can instrument C code to unroll loops or inject additional goals.

Instrumentation can also speed up the verification of programs containing operations over arrays using ghost variables and rewriting rules [56]. The three main differences between our and the mentioned approach are (a) that we transform the task from other specifications to reachability specifications, and the above-mentioned approach transforms programs with extended quantifiers in assertions to programs with simpler assertions, (b) the implicit ordering of instrumentation, and (c) our approach does the matching on a control-flow automaton of the program and not on its syntax.

3   Background

Control-Flow Automata. We model the control-flow of a program as *control-flow automaton (CFA)* [57]. A CFA (L, l_0, G) consists of a finite set L of locations, an initial location $l_0 \in L$, and a set $G \subseteq L \times Ops \times L$ of edges, which represent that the control flows from one location to the next while executing an operation. We use the special function `nondet()`, which returns a non-deterministic value, and the special operation `assert(`π`)`, which means that the condition π over program variables holds whenever the program execution reaches this operation (condition π is a location invariant). Figure 1 shows an example of a program and a corresponding CFA. The program-to-CFA transformation is implemented in the software-verification framework CPACHECKER [8], which we use in TRANSVER as a component. CPACHECKER also stores information about the type of variables and expressions used on the edges of the CFA. We assume that the CFA is constructed from a C program such that every edge contains at most one operation, and at most one arithmetic or boolean expression (composite expressions are decomposed into multiple edges). Since CPACHECKER leaves expressions with multiple arithmetic expressions, e.g., `x + y + z`, during the transformation, we introduce new edges on demand to split the arithmetic operation into, e.g., `tmp = x + y` and `tmp + z`. However, instead of creating explicit auxiliary variables, we keep track of which subexpressions should be substituted in their place. A *program state* is a mapping from program variables to their values, including a program counter `pc`, which is a variable that is mapped to the current program location.

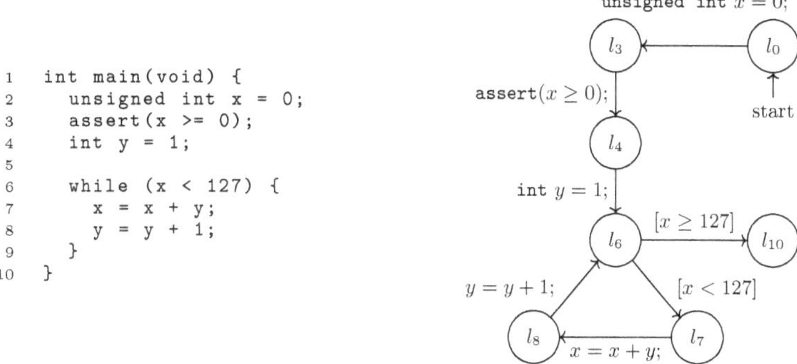


```
1   int main(void) {
2     unsigned int x = 0;
3     assert(x >= 0);
4     int y = 1;
5
6     while (x < 127) {
7       x = x + y;
8       y = y + 1;
9     }
10  }
```

Fig. 1: Example program (left) with corresponding CFA (right)

Table 1: Specifications considered in this work, as defined by SV-COMP

Specification	Explanation
reachability	The function **reach_error** is not called in any execution of the program. We write **assert**(π) for **if (!π) reach_error()**.
no-overflow	No execution of the program produces during an operation a signed integer value that is outside the range of the signed C type.
termination	No execution of the program has infinitely many operations.
memory cleanup	No execution of the program allocates a pointer and then terminates without freeing it.

Specifications. One of the most prominent specification languages for behavioral properties is linear-time temporal logic (LTL) [58]. In the International Competition on Software Verification (SV-COMP) [59], verifiers compete in verifying several practical LTL properties[2]. Table 1 lists the specifications from SV-COMP for which we showcase transformations within our framework in Sect. 5. There are two important subclasses of LTL formulas: *safety* and *liveness*.

Safety specifications describe properties that must hold for all reachable program states on all program executions ("something bad never happens"). A violation consists of a finite execution that has a program state for which the property does not hold. In SV-COMP, the most general version of this kind of specification is *reachability*. Other safety specifications like *no-overflow* and *memory cleanup* can be expressed as reachability by applying a program transformation.

Liveness specifications describe properties that must eventually hold on every program execution ("something good eventually happens"). A violation consists of an infinite execution on which the property never holds or a finite execution that terminates before the property holds. SV-COMP uses only one liveness property, program *termination*, which we also showcase in this paper.

---

[2] https://sv-comp.sosy-lab.org/2024/rules.php

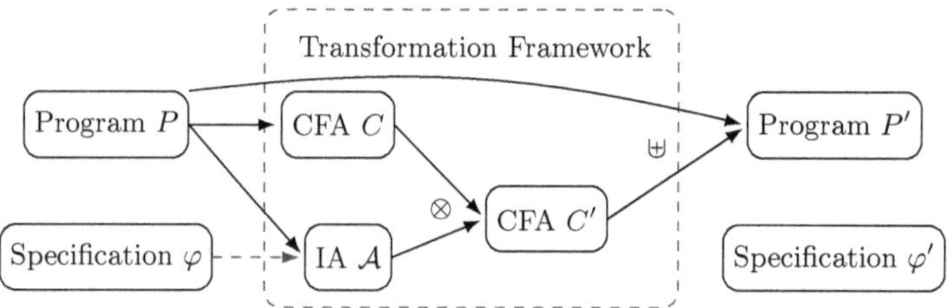

Fig. 2: Workflow of the program transformation using TransVer

4   Transformation Framework

We show the workflow of our transformation framework in Fig. 2. TransVer takes a program P and a specification φ as an input. It uses CPAchecker to construct the corresponding CFA C[3] for program P and combines it with the instrumentation automaton $\mathcal{A}$ to a modified CFA C' using the sequentialization operator $\otimes$. The operations of the new control-flow edges of C', along with the corresponding line numbers at which the operations should be inserted in P, are then exported by CPAchecker. Finally, the exported modifications are added to P using the instrumentation operator $\uplus$, in order to produce the transformed program P'. We use the CFA construction of CPAchecker [3] because it supports a wide variety of C features. For a new specification, the user needs to formalize the transformation as an instrumentation automaton and construct a Java object for it inside CPAchecker, marked by the blue dashed arrow in Fig. 2 from φ to $\mathcal{A}$. The specification can be program-specific or general for a large group of programs. We formalize the automata for three example specifications in Sect. 5.

Instrumentation Automata. An instrumentation automaton specifies how a program needs to be instrumented in order to make the original specification explicit using assertions. It is inspired by the observer automata of the BLAST query language [4]. They allow syntactic pattern matching on a given C program, followed by an action or assertion. To enable the transformation of more complex properties, we additionally match over the structure of the CFA. This makes it possible to match loops that are not immediately apparent from the program's syntax, like gotos.

An *instrumentation automaton (IA)* is a tuple $(Q, q_0, Var, \delta, \alpha)$ that consists of a set Q of states, an initial state q_0, a set Var of automaton variables, a transition relation $\delta \subseteq Q \times PATTERNS \times Ops \times \{A, B\} \times Q$, and a state-annotation function α. A transition (q, ρ, op, X, p) from a state q to a state p specifies:

- A *pattern* $\rho \in PATTERNS$ describes which C expression from a CFA edge is matched. The pattern .* matches any symbols on the CFA edge. Furthermore, loop conditions and their negations can be matched with the special patterns `cond` and `!cond`, while the pattern `true` matches any operation. Moreover, the pattern can also contain capturing groups, and the matching results are

---

[3] In practice, CPAchecker constructs a set of CFAs, one per function. For presentation purposes, and without loss of generality, we use only one CFA in this paper.

assigned to match identifiers $\$x_0, \$x_1, \ldots$, which can be further used like variables in the instrumented operation op. Given a C operation op^C from a CFA edge, $match(\rho, op^C)$ implies that ρ matches op^C and $vars(\rho, op^C)$ is the map of match identifiers $\$x_0, \$x_1, \ldots$ to the operands of op^C. For example, $\rho \equiv . * \$x_0 + \$x_1$; matches $op^C \equiv x + 4$, and $vars(\rho, op^C) = \{x_0 \to x, x_1 \to 4\}$.

- An *operation* $op \in Ops$ is a statement that can read and write to IA variables from Var, but can only read values from the match identifiers and program variables from the CFA. Moreover, it can refer to the match identifiers $\$x_0, \$x_1, \ldots$ that occur in ρ, which $sub(op, m)$ replaces with their mapped expressions from $m \equiv vars(\rho, op^C)$. E.g., for the operation op given by $sum = x_0 + x_1$; with $sum \in Var$, then $sub(op, \{x_0 \to x, x_1 \to 4\})$ results in $sum = x + 4$;.

- The *symbol* $X \in \{A, B\}$ specifies whether the instrumented operation should be placed before (B) or after (A) the matched edge in the CFA.

The state-annotation function $\alpha : Q \to \{true, false, loop\_head, init\}$ assigns to every state of the instrumentation automaton a predicate $L \to \{true, false\}$. The location predicates $true(l)$ and $false(l)$ hold for all locations and no location, respectively, $loop\_head(l)$ holds for program locations l at the beginning of a loop, and $init(l)$ holds for the initial program location l.

Sequentialization Operator. The sequentialization operator $\otimes$ takes the operations from an IA and places them in the indicated locations in an input CFA. The operator implicitly traverses both the CFA and IA in parallel (on-the-fly reduced product). During the traversal, it processes pairs of CFA locations and automaton states (l, q), and pairs of a CFA edge $e = (l, op^C, l')$ and an automaton transition $t = (q, \rho, op, X, p)$. First, it checks whether l matches q with the predicate from the state-annotation function $\alpha(q)$ and whether ρ matches op^C. Afterwards, it instantiates the operation op, substituting the match identifiers from ρ. Lastly, it creates a new edge before or after e, depending on X. If there is no more match for e, the algorithm inserts the original edge e from the input CFA.

The sequentialization operator is implemented as Alg. 1. The algorithm starts with the initialization step, which traverses the CFA and collects additional information about the program that can then be used to construct mulitple concrete instrumentation automata from the input instrumentation automaton for a given program. For example, a transformation for termination initializes one automaton, cf. Fig. 6, per loop in the CFA, collects all the variables used in the respective loop, and initializes one ghost variable for each. Initializing multiple automata is an optimization that users can choose to implement when initializing their defined automaton. Moreover, `initialize_automata` also returns a location from the CFA for each automaton, labeled as the initial location for the sequentialization. This optimization is used to skip parts of the CFA that are irrelevant. For example, it is used by the automaton for termination reduction to monitor one loop per automaton.

Line 2 initializes the sets of locations and the edges of the output CFA. Line 3 instantiates the set `waitlist` to contain the initial pair of program location and IA state, for each initialized IA from line 1. The while loop starting in line 4

Algorithm 1 Sequentialization operator $\otimes$

Input: a CFA $C = (L, l_0, G)$, an IA $\mathcal{A} = (Q, q_0, Var, \delta, \alpha)$
Output: CFA $C' = (L', l_0', G')$
1: $(\mathcal{A}_0, l_{init}^0), \ldots, (\mathcal{A}_k, l_{init}^k) \leftarrow$ `initialize_automata`$(C, \mathcal{A})$;
2: $L', G' \leftarrow \{l_0\}, \{\}$;
3: `waitlist`, `finished` $\leftarrow \{(l_{init}^0, q_0^0), \ldots, (l_{init}^k, q_0^k)\}, \{\}$;
4: **while** `waitlist` $\neq \emptyset$ **do**
5: $\quad (l, q^i) \leftarrow$ `waitlist.pop`$()$;
6: $\quad$ **if** $((l, q^i) \in$ `finished`$)$ **then**
7: $\quad\quad$ `continue`;
8: $\quad$ `finished.add`$((l, q^i))$;
9: $\quad$ `waitlist` $\leftarrow$ `waitlist` $\cup$ `succ`$((l, q^i))$;
10: $\quad$ **if** $\neg\alpha^i(q^i)(l) \vee$ `succ_IA`$((l, q^i)) = \{\}$ **then**
11: $\quad\quad G' \leftarrow G' \cup \{(l, \cdot, \cdot) \in G\}$;
12: $\quad$ **else**
13: $\quad\quad G' \leftarrow G' \cup$ `new_edges`$((l, q^i))$;
14: $\quad L' \leftarrow L' \cup \{l^{new} \mid l^{new} \notin L' \wedge \exists l' : ((l', op, l^{new}) \in G' \vee (l^{new}, op, l') \in G')\}$;
15: **return** (L', l_0', G');

---

traverses both the input CFA and the IA in parallel. States and all the other components from automaton $\mathcal{A}_i$, for $0 \leq i \leq k$, are marked with i in superscript.

In every iteration, it works with a pair (l, q^i) that was not yet processed. The order in which the pairs are processed is not specified. Therefore, users must ensure that the transformation given by their IA does not depend on the exploration order of the waitlist. The algorithm then computes the successors of (l, q^i) as seen in line 9, if it was not yet processed. Function `succ` is defined as follows in Eq. (2).

$$\texttt{succ\_IA}((l, q^i)) = \{(l, p^i) \mid \exists(l, op^C, \cdot) \in G, \tag{1}$$
$$\exists(q^i, \rho, \cdot, \cdot, p^i) \in \delta^i : match(\rho, op^C)\}$$

$$\texttt{succ}((l, q^i)) = \begin{cases} \{(l', q^i) \mid \exists(l, \cdot, l') \in G\} & if \ \neg\alpha^i(q^i)(l) \\ \{(l', q^i) \mid \exists(l, \cdot, l') \in G\} & if \ \texttt{succ\_IA}((l, q^i)) = \{\} \\ \texttt{succ\_IA}((l, q^i)) & otherwise \end{cases} \tag{2}$$

To compute the successors using Eq. (2), we consider three cases. In the first case, the annotation of the state does not hold for the location, and in the second, none of the outgoing edges from q^i matches an outgoing edge from l. Therefore, the algorithm progresses only in the CFA and creates new pairs of q^i with all the successors of l. In the third case, the annotation $\alpha^i(q^i)(l)$ holds, and there are transitions from q^i with a pattern that matches some edge from l. This results in progressing only with the IA and pairing all IA successors of the transitions that matched some edge starting at l, as defined by function `succ_IA` in Eq. (1).

The condition in line 10 guards the addition of new edges into the resulting CFA. If no outgoing edge is matched or the state annotation does not hold for the location, the algorithm adds all the edges from the input CFA C. Otherwise, the function `new_edges` defined in Eq. (4) computes all edges that instrument the original program.

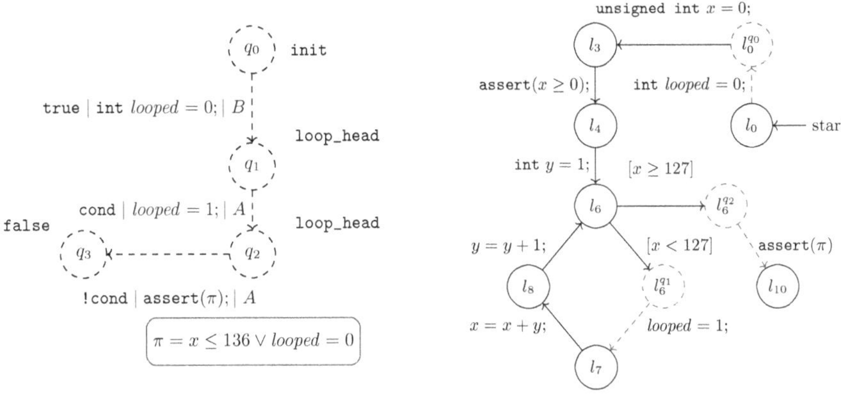

(a) Instrumentation automaton   **(b)** CFA after sequentialization

Fig. 3: Example instrumentation automaton (left) and the example CFA from Fig. 1 after sequentialization with it (right); dashed lines and locations indicate instrumented new CFA edges and locations

The function `new_edges` uses the predicate SO (*Substituted Operation*) presented in Eq. (3). The predicate is true if the operation op^{A_i} is an instantiated operation from a transition in an IA in which the match identifiers $\$x_0, \$x_1, \ldots$ were substituted by the match results (expressions) from op^C. For example, let us assume CFA edge $(l, y = z+42;, l')$ and an IA transition $(q, .*\$x1+\$x2, \texttt{assert}(x1 > x2), B, q')$, then it is the case that $SO(q, y = z+42;, \texttt{assert}(z > 42), B)$ evaluates to true. The function `new_edges` uses the predicate SO to get the operations with replaced variables based on the pattern ρ, in the previous example, it is $\texttt{assert}(z > 42)$. The function places the new operation after (A) or before (B) the edge (separated by location l^{new}) in the original CFA C. Lastly, the algorithm adds the new locations into C' in line 14.

$$SO(q^i, op^C, op^{A_i}, X) =$$
$$\exists (q^i, \rho, \overline{op}^{A_i}, X, \cdot) \in \delta^i : match(\rho, op^C) \wedge op^{A_i} = sub(\overline{op}^{A_i}, vars(\rho, op^C)) \tag{3}$$

$$\texttt{new\_edges}((l, q^i)) =$$
$$\bigcup_{\substack{(x,y,X)\in \\ \{(A_i,C,B), \\ (C,A_i,A)\}}} \left\{ (l, op^x, l^{new}), (l^{new}, op^y, l') \mid (l, op^C, l') \in G \wedge SO(q^i, op^C, op^{A_i}, X) \right\}$$
$$\tag{4}$$

Example 1. Consider the program from Fig. 1 and the property *"If the execution of a program does at least one iteration of the loop, the value of x will always be at most 136"*. The property can be formalized as the instrumentation automaton in Fig. 3a. Applying the sequentialization operator to the IA in Fig. 3a and the CFA in Fig. 1 results in the CFA in Fig. 3b. The function `initialize_automata` returns the same IA as there is nothing to be instantiated in this example. It then initializes `waitlist` with the only pair (l_0, q_0).

$$-,/,*,\dots \quad \boxed{q_0}^{\;\text{true}} \quad .*\$x_0 + \$x_1;\,|\,\mathtt{assert}(\pi_+);\,|\,B$$

$$\pi_+ = \neg\Big((x_1 > 0 \ \wedge\ x_0 > \mathtt{INT\_MAX} - x_1)\ \vee\ (x_1 < 0 \ \wedge\ x_0 < \mathtt{INT\_MIN} - x_1)\Big)$$

Fig. 5: Example instrumentation automaton for the *no-overflow* property

The most important explored pairs are (l_0, q_0), (l_6, q_1), and (l_6, q_2), as they are the only pairs for which the annotation α holds. For example, let us focus on the CFA edge $(l_6, [x < 127], l_7)$ and the IA transition from q_1 to q_2. The predicate $SO(q_1, [x < 127], looped = 1; A)$ is satisfied because the edge matches pattern cond, thus $\mathtt{new\_edges}((l_6, q_1)) = \{(l_6, [x < 127], l_6^{q_1}), (l_6^{q_1}, looped = 1;\,, l_7)\}$.

```
int main(void) {
  int looped = 0;  //
  unsigned int x = 0;
  assert(x >= 0);
  int y = 1;

  while (x < 127) {
    looped = 1;  //
    x = x + y;
    y = y + 1;
  }
  assert(x <= 136 || looped == 0);  //
}
```

Fig. 4: Instrumented program; lines marked with // are inserted operations

Instrumentation Operator. The instrumentation operator ⊎ uses the output CFA from the sequentialization operator and the original program as inputs. It iterates through every statement of the program and in case there are some newly inserted operations in the corresponding CFA location, it adds them before or after the operation in the original program based on their order in the input CFA.

Example 2. Applying the instrumentation operator ⊎ to the CFA in Fig. 3b and the original program in Fig. 1 (left) results in the program in Fig. 4.

5   Specifications as Instrumentation Automata

To study the performance of reachability analyzers when applied to different specifications, we focus our experiments on the transformation of *no-overflow*, *termination*, and *memory cleanup* (as defined in SV-COMP [9]) to reachability. This section shows how to formalize the specifications as instrumentation automata.

No-Overflow. According to SV-COMP, the specification *no-overflow* is violated by a given program if there exists an execution of the program that contains an operation with a signed-integer value that does not fit into the type of the signed integer. The motivation for this specification is that programs should be well-defined, and an overflow of signed integers is an undefined behavior according to the C99 standard [60]. To simplify the presentation, let us assume that a program consists only of signed-integer variables. In our implementation, we track the types of the variables and the expressions on the edges of the CFA, and instrument only the operations with signed-integer results. As explained in Sect. 3

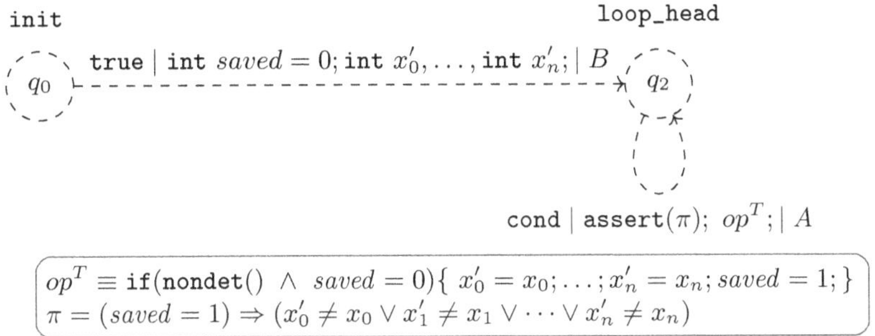

Fig. 6: Example instrumentation automaton for the *termination* property

complex arithmetic expressions are decomposed into multiple CFA edges, each containing one operation. Figure 5 shows the IA corresponding to the no-overflow specification. We display only the transition for addition since the transitions for the other operations are analogous. The automaton matches every operation and adds the corresponding condition as an **assert** before the operation. For example, before an addition, if x_1 is positive, then the result of $x_0 + x_1$ should not be larger than **INT_MAX**, and if x_1 is negative, then the result of $x_0 + x_1$ should not be smaller than **INT_MIN**. The operations together with the necessary conditions to prevent the overflows can be found listed online [61].

Termination. Figure 6 shows the instrumentation automaton for the *termination* property. It is based on a transformation of liveness properties to safety properties for finite systems [62]. To find an infinite execution, the instrumentation monitors the visited states of the execution. If a program state is encountered twice during the execution of a loop, a non-terminating execution has been found. As a preprocessing step, our implementation traverses the whole CFA and initializes an automaton for each loop. For every loop, it collects all variables that are in scope and initializes their shadow copies $x'_0, \ldots, x'_n$. Each time the loop-head is visited, the instrumented program can make a non-deterministic choice to save the state if no state has been saved before. This is performed by the operation op^T. An assertion ensures that if the state was saved previously then the current state is different. A violation of the assertion means that the execution encountered the same state twice, and hence, it can repeat the loop infinitely often. Note that the transformation is complete but not sound, because dynamic structures like linked lists can have unbounded executions without visiting the same state twice.

Memory Cleanup. We assume in this transformation that all memory-allocation functions are called directly and not through a function pointer. Figure 7 shows an IA for this property. It non-deterministically decides to track the pointer being allocated and checks if the tracked pointer has been deallocated when the program terminates. The transition from q_0 to q_1 initializes the tracking of the pointer. In Fig. 7, we draw only one self-loop in q_1 but it actually represents four distinct edges, one for each of the lines above the arrow. The result of every memory-allocating function such as **malloc** and **calloc** is non-deterministically assigned to the tracking pointer **ptr**. For **realloc**, we first check if the pointer being

$$\texttt{free}(\$x_0); \mid \text{if } (ptr == \$x_0) \text{ then } \{ptr = \texttt{NULL}; \} \mid B$$
$$\$x_0 = \texttt{calloc}(\$x_1); \mid \text{if } (\texttt{nondet}()) \text{ then } \{ptr = \$x_0; \} \mid B$$
$$\$x_0 = \texttt{malloc}(\$x_1); \mid \text{if } (\texttt{nondet}()) \text{ then } \{ptr = \$x_0; \} \mid B$$
$$\$x_0 = \texttt{realloc}(\$x_1, \$x_2); \mid \text{if } (ptr == \$x_1) \text{ then } \{ptr = \$x_0; \} \mid B$$

init

q_0 — true | void* ptr; | B → q_1 — abort(); | assert($!ptr$); | B true
return .*; | assert($!ptr$); | B → q_2

true

Fig. 7: Example instrumentation automaton for the *memory-cleanup* property

reallocated is the same as the one being tracked. If it is, we update the tracking pointer. When freeing memory by using `free`, we have to check if the pointer is the one currently being tracked. If it is, we set it to null, i.e., we are currently not tracking any pointer. Finally, when exiting the program, demonstrated through the state q_2, we check if the tracking pointer is null. If not, then there exists at least one execution such that the pointer is not deallocated.

6 Evaluation

The evaluation of our approach addresses the following research questions:

RQ 1 (Modularity): Do transformations enable verifiers that only support reachability to verify properties for which they do not have integrated support?
RQ 2 (Effectiveness): Are state-of-the-art reachability verifiers combined with transformations *effective* compared to state-of-the-art verifiers with integrated specification support?
RQ 3 (Efficiency): Are state-of-the-art reachability verifiers combined with transformations *efficient* compared to state-of-the-art verifiers with integrated specification support?
RQ 4 (Degradation): Is there a performance difference between a verifier using a program transformation on the input program instead of during the analysis?

The proposed research questions divide the evaluation into three parts to provide answers for our initial motivation. First, RQ 1 shows that verifiers supporting only reachability can be adapted to verify other properties without additional engineering effort. Second, RQ 2 and RQ 3 evaluate if the fine-tuned reachability analyses of the best-performing verifiers can be as effective and efficient as state-of-the-art verifiers with integrated specification support. This supports a clear separation of concerns: Boosting the performance of the reachability analysis improves the performance on multiple specifications. Finally, RQ 4 evaluates whether encoding the transformation as a C program results in a performance loss compared to encoding it directly within the verification algorithm.

Transformations for no-overflow and termination do not put any additional requirements on the reachability analyzer. For memory cleanup, the verifiers need to handle memory allocation and deallocation correctly, even after the transformation. The verifiers used as reachability analyzers in RQ 1 do not participate in

Table 2: Verifiers used in the experiments

Verifier	Version	reachability	no-overflow	memory cleanup	termination
2ls [67]	[68]	✓	✗	✗	✓
CPAchecker [69]	[70]	✓	✓	✓	✓
CPA-BAM-SMG [71]	[71]	✓	✓	✓	✓
CPV [72]	[73]	✓	✗	✗	✗
EmergenTheta [12]	[74]	✓	✗	✗	✗
PredatorHP [75]	[76]	✓	✗	✓	✗
Symbiotic [52]	[77]	✓	✓	✓	✓
Theta [13]	[78]	✓	✗	✗	✗
UAutomizer [79]	[80]	✓	✓	✓	✓
UTaipan [81]	[82]	✓	✓	✗	✗

the memory-safety category of SV-COMP, hence, we assume they do not model memory with sufficient precision. Therefore, we answer only RQ 2 and RQ 3 for memory cleanup. We also exclude memory cleanup from RQ 4 since there is no internal transformation for memory cleanup inside CPAchecker. We do not include the time for the transformation in the comparison as it is negligible.

Benchmark Set. To answer the proposed research questions, we use a subset with 2529 tasks for no-overflow, a subset with 779 tasks for termination, and the full set with 41 tasks[4] for memory cleanup of the SV-Benchmarks collection at its SV-COMP 2025 version [63], the largest dataset of C programs with known verification verdicts for several properties. The chosen subsets of the benchmark set do not include programs containing dynamic data structures and arrays for property termination, and do not include programs with recursive function calls for property no-overflow. These programs were not included in the evaluation because our current implementation does not support these program features. If the expected verdict for a verification task is *true* (property holds), we call it a *proof*, if it is *false* (property does not hold), we call it an *alarm*.

Verifiers Evaluated. Table 2 lists all evaluated verifiers together with their supported properties. We selected sound[5] and open-source[6] verifiers that participated in SV-COMP 2024.[7] We add -R to the name of a verifier when used to verify a transformed program (for reachability). For program transformation, we used version 1.0.1 of TransVer, which uses CPAchecker at commit 66247485.

Benchmark Environment. For conducting our evaluation, we use BenchExec to ensure reliable benchmarking [83]. All benchmarks are performed on machines with an Intel Xeon E5-1230 CPU (4 physical cores with 2 processing units each), 33 GB of RAM, and running Ubuntu 24.04 as operating system. Each verification task is limited to 900 s of CPU time, 15 GB of memory, and 1 physical core

---

[4] We excluded the tasks in `Juliet.set` because they were not used in SV-COMP 2024.

[5] Proton [64] won the termination category, but applies unsound approximations.

[6] VeriAbsL [65] and VeriAbs [66], 1st and 2nd place in the reachability category, are not open-source.

[7] We used the 2025 version of CPV, because the 2024 version is based on cgroups v1.

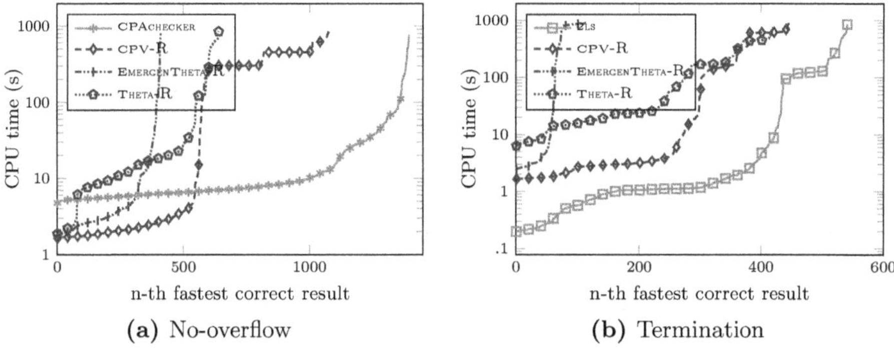

(a) No-overflow   **(b)** Termination

Fig. 8: *RQ 1 (Modularity)*: Quantile plots comparing reachability verifiers (without integrated support for the properties) on the transformed programs with the third-best verifier (with integrated support of the properties) in the respective SV-COMP 2024 category on the original programs

(2 processing units). The limits for time and memory are the same as used in SV-COMP 2024, but the competition used 2 physical cores (4 processing units).

6.1   RQ 1: Modularity

Since TRANSVER produces C programs, it directly allows any verifier for C programs that supports reachability to also analyze other specifications. We consider three verifiers that supported only reachability in SV-COMP 2024, namely CPV, THETA, and EMERGENTHETA. Figures 8a and 8b compare these verifiers with the third-best verifier of SV-COMP 2024 in the categories no-overflow and termination, respectively.

For no-overflow, CPV-R was only 317 tasks behind CPACHECKER, and for termination, CPV-R was only 94 tasks behind 2LS. The verifiers supporting only reachability could solve roughly two-thirds of the tasks on average that the third-best performing verifier could solve in the respective category. Notable is that CPV was 5th, THETA was 18th, and EMERGENTHETA was 20th for reachability in SV-COMP 2024 [9]. This indicates that reachability approaches can be useful for verifying different specifications even if they are outperformed on reachability tasks.

> TRANSVER makes it possible for reachability verifiers to analyze specifications for which they do not have integrated support.

6.2   RQ 2: Effectiveness

To answer RQ 2 and RQ 3, we compare the results for two groups of verifiers: (a) the two best verifiers in the overall category in SV-COMP 2024, CPACHECKER and UAUTOMIZER, on the transformed programs, and (b) the best verifiers in SV-COMP 2024 in the respective category with integrated support for the property (cf. [9, Table 10], also Table 2) on the original tasks.

Table 3: *RQ 2 (Effectiveness)*: Results for 2529 transformed no-overflow tasks

Results (#Tasks)	CPACHECKER	UAUTOMIZER	UTAIPAN	CPACHECKER-R	UAUTOMIZER-R
Correct 2529	1 396	1 852	**1 856**	1 583	1 731
Proofs 2034	1 046	1 450	**1 456**	1 197	1 376
Alarms 495	350	**402**	400	386	355
Incorrect	1	0	0	2	1
Proofs	0	0	0	2	1
Alarms	1	0	0	0	0

Table 4: *RQ 2 (Effectiveness)*: Results for 779 transformed termination tasks

Results (#Tasks)	2LS	UAUTOMIZER	CPACHECKER-R	UAUTOMIZER-R
Correct 779	**541**	411	396	515
Proofs 384	224	268	118	**294**
Alarms 395	**317**	143	278	221

No-Overflow. Table 3 shows that UTAIPAN and UAUTOMIZER were able to provide 1856 (73 %) and 1852 (73 %) correct results, respectively. UAUTOMIZER-R solved 1731 (68 %) tasks correctly. CPACHECKER-R was able to find 151 more proofs and 36 more alarms than CPACHECKER with its integrated no-overflow analysis. We inspected all incorrect results and concluded that they were not caused by our transformation, since the other reachability analyzers were able to solve them correctly. It is expected that reachability and no-overflow algorithms are conceptually similar as they are both safety properties. However, the transformation allows us to use other algorithms for reachability which leads to an increase of the performance for CPACHECKER. In this particular case, the difference is likely due to the overflow analysis of CPACHECKER being based on predicate abstraction [84] and the reachability analysis being a portfolio approach including k-Induction, predicate analysis, and value analysis. With TRANSVER, we can leverage the strength of the portfolio for no-overflow tasks.

Termination. Table 4 shows the results for termination. UAUTOMIZER-R performs better than the termination analysis of UAUTOMIZER. It was able to solve 515 (66 %) of the tasks, and provide 26 more proofs and 78 more alarms than UAUTOMIZER with its integrated termination analysis. In contrast to no-overflow, where the performance gain could be attributed to the used algorithms, the difference in the performance for termination is more likely due to the conceptual differences between the verification approaches, because the algorithms developed to analyze termination are usually different from the algorithms for reachability. Thanks to the transformation framework, we had identified 17 tasks with undefined behavior in the form of signed-integer overflows. They were excluded in SV-COMP 2025.[8]

---

[8] Since the termination behavior is not defined in this case, we removed them from the comparison and from SV-Benchmarks: `https://gitlab.com/sosy-lab/benchmarking/sv-benchmarks/-/merge_requests/1543`

Table 5: *RQ 2 (Effectiveness)*: Results for 41 transformed memory-cleanup tasks

Results (#Tasks)	PredatorHP	Symbiotic	CPA-BAM-SMG	CPAchecker-R	UAutomizer-R
Correct 41	35	**39**	34	33	23
proofs 2	1	**2**	1	0	0
alarms 39	34	**37**	33	33	23

Memory Cleanup. Table 5 shows the results of the comparison of the verifiers on the transformed memory-cleanup tasks. The results show no large discrepancy between the different verifiers. However, it is notable that CPAchecker-R could solve almost as many tasks as CPA-BAM-SMG in this case and that it is close to the performance of PredatorHP which specializes in memory analysis.

Threats to Validity. As mentioned in paragraph "Benchmark Set", the benchmark set we used contains only a subset of the programs from SV-COMP, and hence, the evaluation on the full dataset could provide a different result.

> The experiments show that the combination of the transformations and reachability verifier can outperform state-of-the-art verifiers for the other specifications. For all the properties, we can see that the best performing verifiers are verifiers with integrated support of the property.

6.3   RQ 3: Efficiency

To report the results on efficiency, we show quantile plots for the correct results of the three considered specifications in Fig. 9.

No-Overflow. Figure 9a shows that while CPAchecker-R can solve more tasks than CPAchecker, it uses a bit more CPU time overall. This can be explained by the configuration of CPAchecker: the integrated no-overflow support is based on predicate analysis only, while, thanks to the flexibility gained by the transformation, we now use CPAchecker's sequential portfolio of reachability analyses for CPAchecker-R.

Termination. Figure 9b shows that UAutomizer-R on the transformed programs is more efficient than UAutomizer on the original tasks, solving more tasks in less CPU time overall.

Memory Cleanup. The plot shows that there is no large difference in efficiency between the verifiers, besides about 10 seconds of JVM startup time for UAutomizer and CPAchecker. The efficiency is expected to be slightly worse on transformed programs, because the transformation introduces a non-deterministic choice for each allocation, making the programs more difficult to verify.

> CPAchecker-R tends to be less efficient compared to state-of-the-art verifiers with integrated support of the specifications. For UAutomizer, there is a significant improvement of the efficiency for the termination specification.

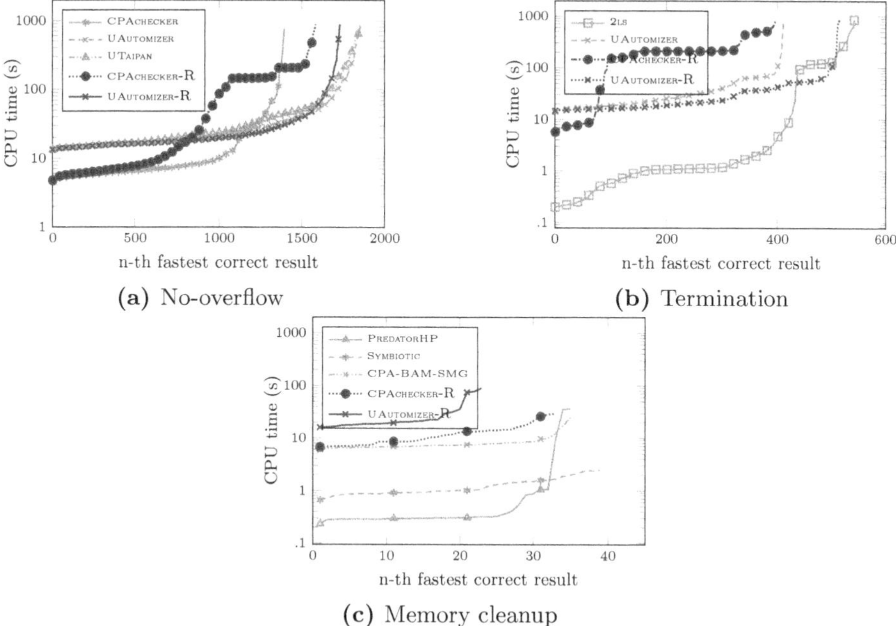

(a) No-overflow

(b) Termination

(c) Memory cleanup

Fig. 9: *RQ 3 (Efficiency)*: Results for verifiers on original and transformed tasks

6.4   RQ 4: No Degradation

Some of the verifiers internally transform various properties to reachability. They usually do it by instrumenting their intermediate representation of an input program or by reflecting the assertion checks in their analyzing algorithm. There are two algorithms in CPAchecker applying the latter. Figure 10 shows the CPU time in seconds for correctly solved tasks by both approaches, using the transformed programs versus using the internal transformation of the property.

For termination (+), both approaches used bounded model checking [85] as the reachability algorithm. It usually solves the task very quickly or does not solve it at all. We can see that for a lot of the tasks, there is a difference between 1-20 seconds if we represent the specification in the input program which is not too much in relation to the 900 seconds time limit. There were many cases where the reachability analysis was faster than the integrated analysis. This is probably due to a better-tuned BMC implementation for reachability.

For no-overflow ( ), both approaches used predicate abstraction [84] as the reachability algorithm. The number of tasks solved by both approaches is much larger than for termination. For most of the tasks, there is no clear overhead in either direction, though there are a few outliers in both directions.

> There is no clear degradation of efficiency using the external transformation for no-overflow. For termination, there is a slow-down in some cases, which is balanced by a speed-up in more other cases. In sum, the modular transformation outside the verifier does not lead to a general degradation of the performance.

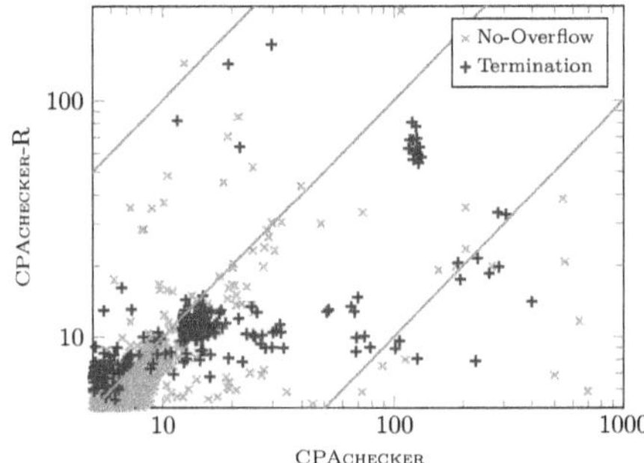

Fig. 10: *RQ 4 (No degradation)*: CPU time of program transformation using internal (CPAchecker) vs. external (TransVer + CPAchecker-R) transformation

7 Conclusion

Developing a verifier for software verification is challenging and requires a large engineering effort. The effort is even larger for supporting various specifications. Verification tools sometimes use internal transformations to mitigate the development time. However, these transformations are usually not modular and have to be developed and maintained separately for every verifier, and for each specification. Our contribution is to offer a new modular framework, implemented in TransVer, which separates the concerns of a high-performance reachability algorithm from supporting other specifications. We demonstrated how the framework works by implementing the transformations for three interesting specifications.

We showed that the construction of new verifiers as a composition of an off-the-shelf transformation and a following off-the-shelf reachability analysis is usually also efficient and effective, and can compete with (and sometimes outperform) state-of-the-art verifiers for no-overflow and termination analysis. Furthermore, the experiments demonstrate that our approach can enable verifiers like CPV or Theta to be competitive in the verification of properties for which they did not have integrated support so far. There was no significant general performance regression on transformed programs compared to using internal transformations.

Data-Availability Statement. A reproduction package containing all data and verifiers used for the experiments is available [86].

Funding Statement. This project was funded by the Deutsche Forschungsgemeinschaft (DFG) — 378803395 (ConVeY) and the Free State of Bavaria.

References

1. Clarke, E.M., Kröning, D., Lerda, F.: A tool for checking ANSI-C programs. In: Proc. TACAS. pp. 168–176. LNCS 2988, Springer (2004). https://doi.org/10.1007/978-3-540-24730-2_15
2. Griggio, A., Jonáš, M.: Kratos2: An SMT-based model checker for imperative programs. In: Proc. CAV. pp. 423–436. Springer (2023). https://doi.org/10.1007/978-3-031-37709-9_20
3. Baier, D., Beyer, D., Chien, P.C., Jakobs, M.C., Jankola, M., Kettl, M., Lee, N.Z., Lemberger, T., Lingsch-Rosenfeld, M., Wachowitz, H., Wendler, P.: Software

verification with CPAchecker 3.0: Tutorial and user guide. In: Proc. FM. pp. 543–570. LNCS 14934, Springer (2024). https://doi.org/10.1007/978-3-031-71177-0_30

4. Beyer, D., Chlipala, A.J., Henzinger, T.A., Jhala, R., Majumdar, R.: The Blast query language for software verification. In: Proc. SAS. pp. 2–18. LNCS 3148, Springer (2004). https://doi.org/10.1007/978-3-540-27864-1_2

5. Ball, T., Rajamani, S.K.: SLIC: A specification language for interface checking (of C). Tech. Rep. MSR-TR-2001-21, Microsoft Research (2002)

6. Ball, T., Rajamani, S.K.: The Slam project: Debugging system software via static analysis. In: Proc. POPL. pp. 1–3. ACM (2002). https://doi.org/10.1145/503272.503274

7. O. Šerý: Enhanced property specification and verification in Blast. In: Proc. FASE. pp. 456–469. LNCS 5503, Springer (2009). https://doi.org/10.1007/978-3-642-00593-0_32

8. Beyer, D., Keremoglu, M.E.: CPAchecker: A tool for configurable software verification. In: Proc. CAV. pp. 184–190. LNCS 6806, Springer (2011). https://doi.org/10.1007/978-3-642-22110-1_16

9. Beyer, D.: State of the art in software verification and witness validation: SV-COMP 2024. In: Proc. TACAS (3). pp. 299–329. LNCS 14572, Springer (2024). https://doi.org/10.1007/978-3-031-57256-2_15

10. Metta, R., Medicherla, R.K., Karmarkar, H.: VeriFuzz: Fuzz centric test generation tool (competition contribution). In: Proc. FASE. pp. 341–346. LNCS 13241, Springer (2022). https://doi.org/10.1007/978-3-030-99429-7_20

11. Metta, R., Medicherla, R.K., Chakraborty, S.: BMC+Fuzz: Efficient and effective test generation. In: Proc. DATE. pp. 1419–1424. IEEE (2022). https://doi.org/10.23919/DATE54114.2022.9774672

12. Bajczi, L., Szekeres, D., Mondok, M., Ádám, Z., Somorjai, M., Telbisz, C., Dobos-Kovács, M., Molnár, V.: EmergenTheta: Verification beyond abstraction refinement (competition contribution). In: Proc. TACAS (3). pp. 371–375. LNCS 14572, Springer (2024). https://doi.org/10.1007/978-3-031-57256-2_23

13. Bajczi, L., Telbisz, C., Somorjai, M., Ádám, Z., Dobos-Kovács, M., Szekeres, D., Mondok, M., Molnár, V.: Theta: Abstraction based techniques for verifying concurrency (competition contribution). In: Proc. TACAS (3). pp. 412–417. LNCS 14572, Springer (2024). https://doi.org/10.1007/978-3-031-57256-2_30

14. Beyer, D., Jankola, M., Lingsch-Rosenfeld, M., Xia, T., Zheng, X.: A modular program-transformation framework for reducing specifications to reachability. arXiv/CoRR **2501**(16310) (January 2025). https://doi.org/10.48550/arXiv.2501.16310

15. Partsch, H., Steinbrüggen, R.: Program transformation systems. ACM Comput. Surv. **15**(3), 199–236 (1983). https://doi.org/10.1145/356914.356917

16. Visser, E.: A survey of strategies in program-transformation systems. In: Proc. WRS. pp. 109–143. ENTCS 57, Elsevier (2001). https://doi.org/10.1016/S1571-0661(04)00270-1

17. Beyer, D., Lee, N.Z.: The transformation game: Joining forces for verification. In: Principles of Verification: Cycling the Probabilistic Landscape. pp. 175–205. LNCS 15262, Springer (2024). https://doi.org/10.1007/978-3-031-75778-5_9

18. Fischer, B., Inverso, O., Parlato, G.: CSeq: A concurrency pre-processor for sequential C verification tools. In: Proc. ASE. pp. 710–713. IEEE (2013). https://doi.org/10.1109/ASE.2013.6693139

19. Inverso, O., Nguyen, T.L., Fischer, B., La Torre, S., Parlato, G.: Lazy-CSeq: A context-bounded model checking tool for multi-threaded C programs. In: Proc. ASE. pp. 807–812. IEEE (2015). https://doi.org/10.1109/ASE.2015.108

20. Necula, G.C., McPeak, S., Rahul, S.P., Weimer, W.: CIL: Intermediate language and tools for analysis and transformation of C programs. In: Proc. CC. pp. 213–228. LNCS 2304, Springer (2002). https://doi.org/10.1007/3-540-45937-5_16
21. Aho, A.V., Sethi, R., Ullman, J.D.: Compilers: Principles, Techniques, and Tools. Addison-Wesley (1986)
22. Alglave, J., Donaldson, A.F., Kröning, D., Tautschnig, M.: Making software verification tools really work. In: Proc. ATVA. pp. 28–42. LNCS 6996, Springer (2011). https://doi.org/10.1007/978-3-642-24372-1_3
23. Beyer, D., Lingsch-Rosenfeld, M., Spiessl, M.: A unifying approach for control-flow-based loop abstraction. In: Proc. SEFM. pp. 3–19. LNCS 13550, Springer (2022). https://doi.org/10.1007/978-3-031-17108-6_1
24. Beyer, D., Lingsch-Rosenfeld, M., Spiessl, M.: CEGAR-PT: A tool for abstraction by program transformation. In: Proc. ASE. pp. 2078–2081. IEEE (2023). https://doi.org/10.1109/ASE56229.2023.00215
25. Jeannet, B., Schrammel, P., Sankaranarayanan, S.: Abstract acceleration of general linear loops. In: Proc. POPL. pp. 529–540. ACM (2014). https://doi.org/10.1145/2535838.2535843
26. Silverman, J., Kincaid, Z.: Loop summarization with rational vector addition systems. In: Proc. CAV, Part 2. pp. 97–115. LNCS 11562, Springer (2019). https://doi.org/10.1007/978-3-030-25543-5_7
27. Frohn, F.: A calculus for modular loop acceleration. In: Proc. TACAS (1). pp. 58–76. LNCS 12078, Springer (2020). https://doi.org/10.1007/978-3-030-45190-5_4
28. Madhukar, K., Wachter, B., Kröning, D., Lewis, M., Srivas, M.K.: Accelerating invariant generation. In: Proc. FMCAD. pp. 105–111. IEEE (2015). https://doi.org/https://doi.org/10.1109/FMCAD.2015.7542259
29. Bodden, E., Hendren, L.: The CLARA framework for hybrid typestate analysis. Softw. Tools Technol. Transf. **14**(3), 307–326 (2012). https://doi.org/10.1007/s10009-010-0183-5
30. Beyer, D., Henzinger, T.A., Jhala, R., Majumdar, R.: Checking memory safety with BLAST. In: Proc. FASE. pp. 2–18. LNCS 3442, Springer (2005). https://doi.org/10.1007/978-3-540-31984-9_2
31. Zhang, Y., Xie, X., Li, Y., Chen, S., Zhang, C., Li, X.: EndWatch: A practical method for detecting non-termination in real-world software. In: Proc. ASE. pp. 686–697 (2023). https://doi.org/10.1109/ASE56229.2023.00061
32. Schuppan, V., Biere, A.: Liveness checking as safety checking for infinite state spaces. Electr. Notes Theor. Comput. Sci. **149**(1), 79–96 (2006). https://doi.org/10.1016/j.entcs.2005.11.018
33. Chalupa, M., Strejček, J., Vitovská, M.: Joint forces for memory safety checking. In: Proc. SPIN. pp. 115–132. Springer (2018). https://doi.org/10.1007/978-3-319-94111-0_7
34. Robles, V., Kosmatov, N., Prevosto, V., Rilling, L., Le Gall, P.: METACSL: Specification and verification of high-level properties. In: Proc. TACAS, Part 1. pp. 358–364. LNCS 11427, Springer (2019). https://doi.org/10.1007/978-3-030-17462-0_22
35. Robles, V., Kosmatov, N., Prevosto, V., Rilling, L., Gall, P.L.: Methodology for specification and verification of high-level requirements with METACSL. In: Proc. FormaliSE. pp. 54–67 (2021). https://doi.org/10.1109/FormaliSE52586.2021.00012
36. Blatter, L., Kosmatov, N., Le Gall, P., Prevosto, V.: RPP: Automatic proof of relational properties by self-composition. In: Proc. TACAS. pp. 391–397 (2017). https://doi.org/10.1007/978-3-662-54577-5_22

37. Blatter, L., Kosmatov, N., Le Gall, P., Prevosto, V., Petiot, G.: Static and dynamic verification of relational properties on self-composed C code. In: Proc. TAP. pp. 44–62 (2018). https://doi.org/10.1007/978-3-319-92994-1_3

38. Harman, M., Hu, L., Hierons, R.M., Wegener, J., Sthamer, H., Baresel, A., Roper, M.: Testability transformation. IEEE Trans. Softw. Eng. **30**(1), 3–16 (2004). https://doi.org/10.1109/TSE.2004.1265732

39. Harman, M.: We need a testability transformation semantics. In: Proc. SEFM. pp. 3–17. LNCS 10886, Springer (2018). https://doi.org/10.1007/978-3-319-92970-5_1

40. Beyer, D., Dangl, M., Dietsch, D., Heizmann, M., Lemberger, T., Tautschnig, M.: Verification witnesses. ACM Trans. Softw. Eng. Methodol. **31**(4), 57:1–57:69 (2022). https://doi.org/10.1145/3477579

41. Ayaziová, P., Beyer, D., Lingsch-Rosenfeld, M., Spiessl, M., Strejček, J.: Software verification witnesses 2.0. In: Proc. SPIN. pp. 184–203. LNCS 14624, Springer (2024). https://doi.org/10.1007/978-3-031-66149-5_11

42. Beyer, D., Spiessl, M.: METAVAL: Witness validation via verification. In: Proc. CAV. pp. 165–177. LNCS 12225, Springer (2020). https://doi.org/10.1007/978-3-030-53291-8_10

43. Necula, G.C.: Proof-carrying code. In: Proc. POPL. pp. 106–119. ACM (1997). https://doi.org/10.1145/263699.263712

44. Julien, S.: E-ACSL: Executable ANSI/ISO C specification language (2022), available at `http://frama-c.com/download/e-acsl/e-acsl.pdf`

45. Necula, G.C., McPeak, S., Weimer, W.: CCURED: Type-safe retrofitting of legacy code. In: Proc. POPL. pp. 128–139. ACM (2002). https://doi.org/10.1145/503272.503286

46. Condit, J., Harren, M., McPeak, S., Necula, G.C., Weimer, W.: CCURED in the real world. In: Proc. PLDI. pp. 232–244. ACM (2003). https://doi.org/10.1145/781131.781157

47. Jakobsson, A., Kosmatov, N., Signoles, J.: Fast as a shadow, expressive as a tree: hybrid memory monitoring for C. In: Proc. SAC. pp. 1765–1772. ACM (2015). https://doi.org/10.1145/2695664.2695815

48. Vorobyov, K., Signoles, J., Kosmatov, N.: Shadow state encoding for efficient monitoring of block-level properties. In: Proc. ISMM. pp. 47–58. ACM (2017). https://doi.org/10.1145/3092255.3092269

49. Kiczales, G., Hilsdale, E., Hugunin, J., Kersten, M., Palm, J., Griswold, W.G.: An overview of AspectJ. In: Proc. ECOOP. pp. 327–353. Springer (2001). https://doi.org/10.1007/3-540-45337-7_18

50. Mahrenholz, D., Spinczyk, O., Schröder-Preikschat, W.: Program instrumentation for debugging and monitoring with AspectC++. In: Proc. ISIRC. pp. 249–256 (2002). https://doi.org/10.1109/ISORC.2002.1003713

51. Vitovská, M., Chalupa, M., Strejček, J.: SBT-Instrumentation: A tool for configurable instrumentation of LLVM bitcode. arXiv/CoRR **1810**(12617) (2018). https://doi.org/10.48550/arXiv.1810.12617

52. Jonáš, M., Kumor, K., Novák, J., Sedláček, J., Trtík, M., Zaoral, L., Ayaziová, P., Strejček, J.: SYMBIOTIC 10: Lazy memory initialization and compact symbolic execution (competition contribution). In: Proc. TACAS (3). pp. 406–411. LNCS 14572, Springer (2024). https://doi.org/10.1007/978-3-031-57256-2_29

53. Slabý, J., Strejček, J., Trtík, M.: Checking properties described by state machines: On synergy of instrumentation, slicing, and symbolic execution. In: Proc. FMICS. pp. 207–221. LNCS 7437, Springer (2012). https://doi.org/10.1007/978-3-642-32469-7_14

54. Menezes, R., Aldughaim, M., Farias, B., Li, X., Manino, E., Shmarov, F., Song, K., Brauße, F., Gadelha, M.R., Tihanyi, N., Korovin, K., Cordeiro, L.: ESBMC v7.4: Harnessing the power of intervals (competition contribution). In: Proc. TACAS (3). pp. 376–380. LNCS 14572, Springer (2024). https://doi.org/10.1007/978-3-031-57256-2_24

55. Gadelha, M.R., Monteiro, F.R., Morse, J., Cordeiro, L.C., Fischer, B., Nicole, D.A.: ESBMC 5.0: An industrial-strength C model checker. In: Proc. ASE. pp. 888–891. ACM (2018). https://doi.org/10.1145/3238147.3240481

56. Amilon, J., Esen, Z., Gurov, D., Lidström, C., Rümmer, P.: Automatic program instrumentation for automatic verification. In: Proc. CAV. pp. 281–304 (2023). https://doi.org/10.1007/978-3-031-37709-9_14

57. Beyer, D., Gulwani, S., Schmidt, D.: Combining model checking and dataflow analysis. In: Handbook of Model Checking, pp. 493–540. Springer (2018). https://doi.org/10.1007/978-3-319-10575-8_16

58. Piterman, N., Pnueli, A.: Temporal logic and fair discrete systems. In: Handbook of Model Checking, pp. 27–73. Springer (2018). https://doi.org/10.1007/978-3-319-10575-8_2

59. Beyer, D., Strejček, J.: Improvements in software verification and witness validation: SV-COMP 2025. In: Proc. TACAS (3). pp. 151–186. LNCS 15698, Springer (2025). https://doi.org/10.1007/978-3-031-90660-2_9

60. American National Standards Institute: ANSI/ISO/IEC 9899-1999: Programming Languages — C. American National Standards Institute, 1430 Broadway, New York, NY 10018, USA (1999)

61. INT32-C. Ensure that operations on signed integers do not result in overflow. `https://wiki.sei.cmu.edu/confluence/display/c/INT32-C.+Ensure+that+operations+on+signed+integers+do+not+result+in+overflow`, [Accessed 2025-07-13]

62. Biere, A., Artho, C., Schuppan, V.: Liveness checking as safety checking. In: Proc. FMICS. pp. 160–177. No. 2 in ENTSC 66, Elsevier (2002). https://doi.org/10.1016/S1571-0661(04)80410-9

63. Beyer, D., Strejček, J.: SV-Benchmarks: Benchmark set for software verification (SV-COMP 2025). Zenodo (2025). https://doi.org/10.5281/zenodo.15012096

64. Metta, R., Karmarkar, H., Madhukar, K., Venkatesh, R., Chakraborty, S.: PROTON: Probes for non-termination and termination (competition contribution). In: Proc. TACAS (3). pp. 393–398. LNCS 14572, Springer (2024). https://doi.org/10.1007/978-3-031-57256-2_27

65. Darke, P., Chimdyalwar, B., Agrawal, S., Venkatesh, R., Chakraborty, S., Kumar, S.: VERIABSL: Scalable verification by abstraction and strategy prediction (competition contribution). In: Proc. TACAS (2). pp. 588–593. LNCS 13994, Springer (2023). https://doi.org/10.1007/978-3-031-30820-8_41

66. Afzal, M., Asia, A., Chauhan, A., Chimdyalwar, B., Darke, P., Datar, A., Kumar, S., Venkatesh, R.: VERIABS: Verification by abstraction and test generation. In: Proc. ASE. pp. 1138–1141. IEEE (2019). https://doi.org/10.1109/ASE.2019.00121

67. Malík, V., Schrammel, P., Vojnar, T., Nečas, F.: 2LS: Arrays and loop unwinding (competition contribution). In: Proc. TACAS (2). pp. 529–534. LNCS 13994, Springer (2023). https://doi.org/10.1007/978-3-031-30820-8_31

68. Viktor, M., František, N., Schrammel, P., Vojnar, T.: 2LS. Zenodo (2023). https://doi.org/10.5281/zenodo.10184626

69. Baier, D., Beyer, D., Chien, P.C., Jankola, M., Kettl, M., Lee, N.Z., Lemberger, T., Lingsch-Rosenfeld, M., Spiessl, M., Wachowitz, H., Wendler, P.: CPACHECKER 2.3

with strategy selection (competition contribution). In: Proc. TACAS (3). pp. 359–364. LNCS 14572, Springer (2024). https://doi.org/10.1007/978-3-031-57256-2_21

70. Beyer, D., Wendler, P.: CPAchecker release 2.3 (unix). Zenodo (2023). https://doi.org/10.5281/zenodo.10203297

71. Vasilyev, A.: CPA-BAM-SMG (SV-COMP 2023). Zenodo (2022). https://doi.org/10.5281/zenodo.10396261

72. Chien, P.C., Lee, N.Z.: CPV: A circuit-based program verifier (competition contribution). In: Proc. TACAS (3). pp. 365–370. LNCS 14572, Springer (2024). https://doi.org/10.1007/978-3-031-57256-2_22

73. Chien, P.C., Lee, N.Z.: CPV release 0.6. Zenodo (2024). https://doi.org/10.5281/zenodo.14203582

74. Bajczi, L., Szekeres, D., Mondok, M., Molnár, V.: EmergenTheta - SV-COMP'24 verifier archive. Zenodo (2023). https://doi.org/10.5281/zenodo.10198872

75. Peringer, P., Šoková, V., Vojnar, T.: PredatorHP revamped (not only) for interval-sized memory regions and memory reallocation (competition contribution). In: Proc. TACAS (2). pp. 408–412. LNCS 12079, Springer (2020). https://doi.org/10.1007/978-3-030-45237-7_30

76. Šoková, V., Peringer, P., Vojnar, T., Kinšt, O.: PredatorHP. Zenodo (2023). https://doi.org/10.5281/zenodo.10183805

77. Jonáš, M., Kumor, K., Novák, J., Sedláček, J., Shandilya, S., Trtík, M., Zaoral, L., Strejček, J.: Symbiotic 10: Submission to SV-COMP 2024. Zenodo (2023). https://doi.org/10.5281/zenodo.10202594

78. Bajczi, L., Telbisz, C., Somorjai, M., Ádám, Z., Dobos-Kovács, M., Szekeres, D., Molnár, V.: Theta - SV-COMP'24 verifier archive. Zenodo (2023). https://doi.org/10.5281/zenodo.10202679

79. Heizmann, M., Bentele, M., Dietsch, D., Jiang, X., Klumpp, D., Schüssele, F., Podelski, A.: Ultimate Automizer and the abstraction of bitwise operations (competition contribution). In: Proc. TACAS (3). pp. 418–423. LNCS 14572, Springer (2024). https://doi.org/10.1007/978-3-031-57256-2_31

80. Dietsch, D., Bentele, M., Heizmann, M., Klumpp, D., Podelski, A., Schüssele, F.: Ultimate Automizer SV-COMP 2024. Zenodo (2023). https://doi.org/10.5281/zenodo.10203545

81. Dietsch, D., Heizmann, M., Klumpp, D., Schüssele, F., Podelski, A.: Ultimate Taipan 2023 (competition contribution). In: Proc. TACAS (2). pp. 582–587. LNCS 13994, Springer (2023). https://doi.org/10.1007/978-3-031-30820-8_40

82. Dietsch, D., Bentele, M., Heizmann, M., Klumpp, D., Podelski, A., Schüssele, F.: Ultimate Taipan SV-COMP 2024. Zenodo (2023). https://doi.org/10.5281/zenodo.10203547

83. Beyer, D., Löwe, S., Wendler, P.: Reliable benchmarking: Requirements and solutions. Int. J. Softw. Tools Technol. Transfer **21**(1), 1–29 (2019). https://doi.org/10.1007/s10009-017-0469-y

84. Beyer, D., Dangl, M., Wendler, P.: A unifying view on SMT-based software verification. J. Autom. Reasoning **60**(3), 299–335 (2018). https://doi.org/10.1007/s10817-017-9432-6

85. Biere, A., Cimatti, A., Clarke, E.M., Strichman, O., Zhu, Y.: Bounded model checking. Advances in Computers **58**, 117–148 (2003). https://doi.org/10.1016/S0065-2458(03)58003-2

86. Beyer, D., Jankola, M., Lingsch-Rosenfeld, M., Xia, T., Zheng, X.: Reproduction package for SPIN 2025 article 'TransVer: A modular program-transformation framework for reduction to reachability'. Zenodo (2025). https://doi.org/10.5281/zenodo.15833440

(Asynchronous) Temporal Logics for Hyperproperties on Finite Traces

Alberto Bombardelli[1,2]([✉])[iD], Laura Bozzelli[3][iD], Cesar Sanchez[4][iD], and Stefano Tonetta[1][iD]

[1] Fondazione Bruno Kessler, Via Sommarive, 18, 38123 Povo, TN, Italy
`abombardelli@fbk.eu`
[2] University of Trento, Via Sommarive, 9, 38123 Povo, TN, Italy
[3] University of Napoli "Federico II", Napoli, Italy
[4] IMDEA Software Institute, Pozuelo de Alarcón, Madrid, Spain

Abstract. Hyperproperties are properties of systems that relate more than one trace, which are common in information-flow, concurrency, symmetry, diagnosability, etc. Several temporal logics for hyperproperties have been proposed—like HyperLTL and HyperCTL* where traces are compared synchronously—and recently extended for comparing traces asynchronously. However, all these temporal hyperlogics are for infinite traces, in spite of the recent popularity of temporal logics on finite traces.

In this paper, we introduce SC-HyperLTL, a logic for asynchronous hyperproperties on *finite traces* and *with past operators*, by adapting and combining two existing logics for asynchronous hyperproperties: stuttering HyperLTL (HyperLTL$_S$) and context HyperLTL (HyperLTL$_C$). We show that SC-HyperLTL is decidable and can encode many hyperproperties of interest. We present two practical algorithms that reduce the model checking problem of SC-HyperLTL to inputs of two existing tools: AutoHyper (which can model check hyperproperties on infinite traces) and nuXmv (a model checker for trace properties). We illustrate our algorithms on a prototype implementation.

1 Introduction

Hyperproperties [18] are properties of systems that relate more than one trace at a time, in contrast with trace properties whose semantics are defined on a single trace. Hyperproperties are common in information-flow security [26,38], robustness and sensitivity of cyber-physical systems [37], concurrency [12] and symmetry [23]. HyperLTL [17] is a temporal logic for hyperproperties that equips LTL with quantifiers for different traces. HyperLTL includes relational capabilities to compare different traces and temporal modalities where the temporal operators move all traces synchronously. Several algorithms and verification methods for HyperLTL have been proposed [20,23,30].

In settings like software verification execution traces are finite and of varying length, and relevant events happen at different instants. Therefore, the semantics of the temporal operators need to cope with when traversing the different

G. Ernst and K. Y. Rozier (Eds.): SPIN 2025, LNCS 15945, pp. 25–43, 2026.
https://doi.org/10.1007/978-3-032-06847-7_2

traces at different speeds. Several recent papers [1,3,14,15,28] have introduced hyperlogics to capture asynchrony between the traces. However, all these temporal logics for hyperproperties, synchronous and asynchronous, consider only *infinite traces* and do not have *past operators*.

At the same time, there is a growing interest in linear temporal logics for finite traces [22,25,32] because of their applications to AI, synthesis and runtime verification. An obstacle for a temporal hyper-logic on finite traces is that, even though the beginning of the trace is a point of synchrony between all traces, the end of the trace may arrive at different instants. This is similar when traversing the trace backwards with past operators.

In this paper we introduce SC-HyperLTL, a novel temporal logic for hyperproperties on finite traces that combines modalities from two known asynchronous logics [14,15] for infinite traces: stuttering HyperLTL (HyperLTL$_S$) and context HyperLTL (HyperLTL$_C$). HyperLTL$_S$ allows to stutter on each trace individually and define points of evaluation that can be arbitrarily apart, while HyperLTL$_C$ allows to restrict which traces proceed according to the temporal operators. We prove that SC-HyperLTL can express many properties of interest and its model-checking is decidable. Then, we present two approaches for the practical model checking $M \models \varphi$ of our logic.

1. We first show a stuttering translation that modifies M to account for individual trace stuttering, and introduces additional constraints to φ to align the traces. The result is a HyperLTL model-checking problem whose output correspond to $M \models \varphi$. This approach only works for quantifier alternation free formulas.
2. The second algorithm introduces k history variables and equips the formula with constraints—local to each trace—for the correctness of the auxiliary variables. This approach works for arbitrary quantifier alternation, and can verify systems where sequences of k observations, which can be arbitrarily far apart, are sufficient to prove or disprove the hyperproperty.

In summary, the contributions of this paper are: (1) A novel logic SC-HyperLTL for (asynchronous) hyperproperties on finite traces with past operators, introduced in Sect. 3. (2) Two practical model checking algorithms for SC-HyperLTL, presented in Sect. 4. (3) A report on a proof-of-concept implementation of these algorithms, reported in Sect. 5.

Motivating Example. Consider the following terminating function P, with input *in* and output *out* which iteratively set *out* to $10(in + 1)$.

```
1    int tmp = 1;
2    out = in;
3    for (int i=0;i<10;i++){
4        int j = 0;
5        while (j++ < in) tmp++;
6        out = tmp;
7    }
```

We would like to show that—in spite of the internal program dynamics—if in is smaller at the beginning of the execution, the n-th output will be also smaller during the whole execution. We call this property *reactive monotonicity*, which requires to relate multiple executions of the same program and, since the program has to terminate, to reason with finite semantics. One attempt using a variation of HyperLTL with finite semantics is:

$$\varphi_{rmon}^{sync} := \forall x.\forall y.in[x] < in[y] \rightarrow \mathbf{G}(out[x] < out[y])$$

which expresses that, for each pair of executions of P (here represented by trace variables x and y) if x starts with an input lower than that of y ($in[x] < in[y]$), then it is always the case that the output of x is smaller than the output of y: $\mathbf{G}(out[x] < out[y])$. During its execution, the program might perform internal computations that are not relevant to the desired property but that can change the outcome due to the synchronous alignment. Therefore, even if P guarantees that the same assignment sequences of the variable out are "reactive monotonic" in two executions, the property can be violated:

$$\sigma_1 := \cdots \rightarrow (j = 2, l = 5, out = 2, in = 2) \rightarrow (j = 3, l = 6, out = 1) \rightarrow (out = 3)$$
$$\sigma_2 := \cdots \rightarrow (j = 2, l = 5, out = 3, in = 3) \rightarrow (j = 3, l = 5, out = 1) \rightarrow (out = 3)$$

To fix this excessive synchrony, we adopt a semantics that evaluates the traces *asynchronously* in general but *synchronously* over a set of interesting points. In our example, we want to check the values of out at those points in which out changes, which is achieved with the *stuttering* variants of the temporal modalities. We also introduce *context* temporal modalities to reason over a subset of the traces in a formula, restricting the temporal progress to those traces. The use of contexts in this example is also motivated by the additional challenge that finite traces introduce. The resulting formula is:

$$\varphi_{rmon}^{SC} := \forall x.\forall y.\{out\}.in[x] < in[y] \rightarrow \mathbf{G}((end[x] \leftrightarrow end[y]) \wedge out[x] < out[y])$$

where $end[x]$ captures whether trace x has a successor, defined as $\langle x \rangle \mathbf{X}\top$ using the context modality. The set $\{out\}$ denotes that the interesting positions are the points in which out changes. The globally modality is then evaluated only over interesting positions i.e. where out changes. Finally, φ_{rmon}^{SC} uses singleton contexts with next operator in $end[x]$ and $end[y]$ to ensure that the traces have the same amount of "interesting points", by requiring that x has a successor iff y has a successor.

Related Work. Starting from HyperLTL and HyperCTL* [17], many other decidable temporal logics for hyperproperties have been proposed, including HyperQPTL [19,35] and HyperPDL-Δ [27]. The semantics of all these logics are synchronous, meaning that temporal modalities move all traces in a lockstep manner. Moreover, none of these logics consider past operators.

Recently, several works have introduced hyperlogics capable of handling an asynchronous traversal of traces, which is useful for addressing speed variations introduced by compiler optimizations, scheduling, and other factors. In [28], the temporal fixpoint calculus H_μ is introduced, featuring operators to describe the independent progress of each trace. In [5], the asynchronous hyperlogic Observation HyperLTL is introduced. This logic is decidable and similar to the one defined here. The main differences between the two are that (i) SC-HyperLTL allows reasoning locally on traces using singleton contexts, and (ii) observational HyperLTL considers different observation conditions for each trace, whereas our logic selects a set of LTL formulas to define identical observations for all traces. Similarly, [3] presents *Asynchronous* HyperLTL (A-HLTL), which extends HyperLTL with the notion of trajectories, introduced in [11], to control the progress of the traces. Our first algorithm is similar to the stutter algorithm from [3], but our fragment is strictly richer, allowing synchronous bounded operators as well and directly supporting past operators. Two other extensions of HyperLTL [14] are stuttering HyperLTL (HyperLTL$_S$), which describes with LTL formulas the interesting points of each trace, and context HyperLTL (HyperLTL$_C$), which restricts the progress of specific subsets of traces. In a recent work [8], we defined a general combination of HyperLTL$_S$ and HyperLTL$_C$ for infinite traces with non-prenex quantifiers. The logic presented in that work subsumes the logic considered in this paper. However, that work was purely theoretical, focusing on a general logic and a decidable fragment to handle complex diagnosability specifications such as Finite Delay Diagnosability [13]. All of these logics are defined for infinite traces and most of them do not support past operators. Two exception for finite traces are [24] and [2]. For what regards [24], the logic introduced there is synchronous, lacks past operators, and provides no tool. On the other hand, in [2] the formalism considers both finite traces and asynchronous systems. However, the formalism is conceptually different and based on automata rather than temporal logics.

Several tools have been proposed for verifying synchronous hyperproperties: MCHyper [23] (which reduces alternating-free verification to LTL model checking), AutoHyper [6] (which implements explicit-state automata constructions), and HyperQB [30] (which follows a bounded-model approach by generating queries for a QBF solver). Recently, new works have focused on verifying asynchronous hyperlogics [3–5,21,29]. In [29], the BMC-based verification of HyperLTL has been adapted to handle A-HLTL. While similar in spirit to our second algorithm, the key difference lies in our algorithm's ability to reason over unbounded executions. Another notable difference between the two approaches is in their reduction techniques. Our method reduces the problem to a synchronous HyperLTL model-checking problem, whereas [29] reduces it to a QBF verification problem. The remaining works employ very different techniques. In [5], the authors reduce Observation HyperLTL model checking to a game-based verification using predicate abstraction. In [21], symbolic execution is used to efficiently find counterexamples to asynchronous $\forall\exists$ hyperproperties. Finally, in [4], Hoare logic is used to reason over $\forall\exists$ temporal safety properties in asynchronous programs.

2   Preliminaries

We use $\mathbb{N}$ for the set of natural numbers, (i,j) for the set $\{k \mid i < k < j\}$, $[i,j]$ for $\{k \mid i \leq k \leq j\}$, $[i,j)$ for $\{k \mid i \leq k < j\}$ and $(i,j]$ for $\{k \mid i < k \leq j\}$. We fix a finite set AP of atomic propositions and use $\Sigma = 2^{\mathsf{AP}}$ for the alphabet. A finite trace is a finite sequence $\sigma = a_0 a_1 \cdots$ of letters from Σ is (i.e. a word from Σ^*). Similarly, an infinite trace σ is a word from Σ^ω. Given a finite or infinite word σ, we use $\sigma(i)$ for a_i and σ^i for the suffix $a_i a_{i+1} \cdots$. A *pointed trace* is a pair (σ, i), where $i \in \mathbb{N}$ is a natural number (called the *pointer*). Given a pointed trace (σ, i) and $n \in \mathbb{N}$, we use $(\sigma, i) + n$ for $(\sigma, i + n)$.

Symbolic Transition System. A *Symbolic Transition System* (STS) M is a tuple $M = \langle V, I, T, F \rangle$ where V is a set of Boolean (state) variables, $I(V)$ is a formula representing the initial states, $T(V, V')$ is a formula representing the transitions, and F is a set of formulae $f(V)$ representing the fairness conditions (final condition on finite traces) i.e. a condition that must occur infinitely often in an infinite trace (at the end of finite traces). A *state* of M is an assignment to the variables V. A *trace* of M is an infinite sequence $s_0, s_1, \ldots$ of states such that $s_0 \models I$, for all $i \geq 0$, $s_i, s'_{i+1} \models T$, and for all $f \in F$, for all $i \geq 0$ there exists $j \geq i$ s.t. $s_j \models f$. Similarly, a finite trace $s_0, s_1, \ldots$ is a finite sequence such that $s_0 \models I$, for all $i \geq 0$, $s_i, s'_{i+1} \models T$, and for all $f \in F$: $(\sigma, |\sigma| - 1) \models f$. Given two transitions systems $M_1 = \langle V_1, I_1, T_1, F_1 \rangle$ and $M_2 = \langle V_2, I_2, T_2, F_2 \rangle$, we denote with $M_1 \times M_2$ the synchronous product $\langle V_1 \cup V_2, I_1 \wedge I_2, T_1 \wedge T_2, F_1 \cup F_2 \rangle$. We refer to $\mathsf{Traces}(M)$ as the set of all infinite traces of M and $\mathsf{Traces}^{fin}(M)$ for the set of non-empty finite traces of M.

The Temporal Logics LTL and LTL$_f$. We consider LTL [34] with past [31] (called PLTL), which has the following syntax:

$$\varphi ::= \top \mid a \mid \varphi \vee \varphi \mid \neg\varphi \mid \mathbf{X}\varphi \mid \mathbf{Y}\varphi \mid \varphi \, \mathbf{U} \, \varphi \mid \varphi \, \mathbf{S} \, \varphi$$

where a is a proposition, $\vee$ and $\neg$ are the usual Boolean disjunction and negation, while $\mathbf{X}$, $\mathbf{U}$, $\mathbf{Y}$ and $\mathbf{S}$ are respectively the "next", "until", "yesterday" and "since" temporal operators. The semantics of PLTL associates pointed traces with formulae as follows (we omit Boolean connectives, which are standard):

$$
\begin{aligned}
(\sigma, i) &\models a && \Leftrightarrow a \in \sigma(i) \\
(\sigma, i) &\models \mathbf{X}\varphi && \Leftrightarrow i < |\sigma| - 1 \text{ and } (\sigma, i+1) \models \varphi \\
(\sigma, i) &\models \mathbf{Y}\varphi && \Leftrightarrow i > 0 \text{ and } (\sigma, i-1) \models \varphi \\
(\sigma, i) &\models \varphi_1 \, \mathbf{U} \, \varphi_2 && \Leftrightarrow \text{for some } j \in [i, |\sigma|), (\sigma, j) \models \varphi_2 \text{ and for all } k \in [i, j), (\sigma, k) \models \varphi_1 \\
(\sigma, i) &\models \varphi_1 \, \mathbf{S} \, \varphi_2 && \Leftrightarrow \text{for some } j \in [0, i], (\sigma, j) \models \varphi_2 \text{ and for all } k \in (j, i], (\sigma, k) \models \varphi_1
\end{aligned}
$$

We also use common derived operators like $\bot$, $\wedge$, $\mathbf{R}$ (dual w.r.t. negation of $\mathbf{U}$), $\mathbf{F}$, $\mathbf{G}$, $\mathbf{O}$ ($\mathbf{O}\phi := \top \, \mathbf{S} \, \phi$), $\mathbf{H}$ (dual w.r.t. negation of $\mathbf{O}$), $\mathbf{T}$ (dual w.r.t. negation of $\mathbf{S}$) and $\mathbf{Z}$ (the dual w.r.t. negation of $\mathbf{Y}$). The variant LTL$_f$ of PLTL for finite traces [25,33] borrows the syntax from LTL, with an additional abbreviation: $\mathbf{N}\varphi = \neg\mathbf{X}\neg\varphi$. A Symbolic Transition System M is a model of an PLTL formula φ, denoted by $M \models \varphi$, whenever $\mathsf{Traces}(M) \models \varphi$. Similarly, M models an LTL$_f$ formula φ whenever $\mathsf{Traces}^{fin}(M) \models \varphi$.

The Temporal Logic HyperLTL. HyperLTL [17] is a temporal logic for hyperproperties. The syntax of HyperLTL is:

$$\alpha ::= \exists x.\alpha \mid \forall x.\alpha \mid \varphi \qquad\qquad \varphi ::= a[x] \mid \varphi \vee \varphi \mid \neg\varphi \mid \mathbf{X}\varphi \mid \varphi \, \mathbf{U} \, \varphi$$

where x is a *trace variable* from an infinite set VAR of trace variables. The intended meaning of $a[x]$ is that proposition a holds at the current time in trace x. Trace quantifiers $\exists x$ and $\forall x$ allow reasoning simultaneously about different traces of the computation. Atomic predicates $a[x]$ refer to a single trace x. Given an PLTL formula α, we denote by $\alpha[x]$ the formula obtained by substituting every atomic proposition a with $a[x]$.

Given a HyperLTL formula φ, we use $\mathsf{TrVars}(\varphi)$ for the set of trace variables quantified in φ. A formula φ is well-formed if for all atoms $a[x]$, x is quantified in φ. Given a set of infinite traces T, the semantics of a HyperLTL formula α is defined in terms of *pointed trace assignments*, which are partial mappings of the form $\Pi : \mathsf{TrVars}(\alpha) \rightharpoonup (T \times \mathbb{N})$. We use $Dom(\Pi)$ for the subset of $\mathsf{TrVars}(\varphi)$ for which Π is defined. Given a trace assignment Π, a trace variable x, a trace σ and a pointer p, we denote by $\Pi[x \mapsto (\sigma, p)]$ the assignment that coincides with Π for every trace variable except for x, which is mapped to (σ, p). The trace assignment with empty domain is denoted by Π_0. Also, we use $\Pi + n$ to denote trace assignment Π' such that $\Pi'(x) = \Pi(x) + n$ for all $x \in Dom(\Pi) = Dom(\Pi')$. Given a partial trace assignment Π and a set of traces T, the semantics of HyperLTL is as follows (we again omit $\vee$ and $\neg$):

$$
\begin{aligned}
\Pi \models_T \exists x.\alpha \quad &\Leftrightarrow \text{ for some } \sigma \in T,\ \Pi[x \mapsto (\sigma, 0)] \models_T \alpha \\
\Pi \models_T \forall x.\alpha \quad &\Leftrightarrow \text{ for all } \sigma \in T,\ \Pi[x \mapsto (\sigma, 0)] \models_T \alpha \\
\Pi \models_T \varphi \qquad &\Leftrightarrow \Pi \models \varphi \\
\Pi \models a[x] \qquad &\Leftrightarrow a \in \sigma(p),\ \text{where } (\sigma, p) = \Pi(x) \\
\Pi \models \varphi_1 \, \mathbf{U} \, \varphi_2 &\Leftrightarrow \text{ for some } j \geq 0\ (\Pi + j) \models \varphi_2, \text{and forall } i < j, (\Pi + i) \models \varphi_1 \\
\Pi \models \mathbf{X}\varphi \qquad &\Leftrightarrow (\Pi + 1) \models \varphi
\end{aligned}
$$

Note that quantifiers assign traces to trace variables and set the pointer to the initial position 0. We say that a set of infinite traces T is a model of a HyperLTL formula φ, denoted $T \models \varphi$ whenever $\Pi_0 \models_T \varphi$. We say that a Symbolic Transition System M is a model of a HyperLTL formula φ, denoted by $M \models \varphi$, whenever $\mathsf{Traces}(M) \models \varphi$.

3   The Logic SC-HyperLTL for Finite Traces

We now introduce SC-HyperLTL, a decidable logic that combines the stuttering feature from Stuttering HyperLTL with a limited version of the context feature from Context HyperLTL. Our logic is expressive and at the same time enjoys a decidable model-checking problem. Since both $\mathrm{HyperLTL}_S$ and $\mathrm{HyperLTL}_C$ are undecidable so is their unrestricted combination (see [8]), so our logic can be seen as a (decidable) useful fragment of such a rich logic. This logics extends

HyperLTL with two new features from HyperLTL$_S$ and HyperLTL$_C$: Γ operator and the singleton operator $\langle x \rangle$. The singleton context operator evaluates a formula over a single trace variable x ignoring whether or not the other traces terminated. The Γ-operator is used to restrict the evaluation of formulae to observation points determined by Γ; then, the temporal operators are synchronously evaluated on these observation points.

The syntax of SC-HyperLTL is:

$$\alpha := \forall x.\alpha \mid \exists x.\alpha \mid \Gamma.\varphi$$
$$\varphi := \langle x \rangle \beta[x] \mid a[x] \mid \neg\varphi \mid \varphi \vee \varphi \mid \mathbf{X}\varphi \mid \varphi \, \mathbf{U} \, \varphi \mid \mathbf{Y}\varphi \mid \varphi \, \mathbf{S} \, \varphi$$

where $a \in V$, Γ is a set of PLTL formulae over V, β is a PLTL formula over V and $\beta[x]$ is defined by replacing each variable $a \in V$ occurring in β with $a[x]$.

To define the semantics, we first introduce Γ-stutter factorizations, adapting the definition from [14].

Definition 1 (Γ-stutter factorization). *Let Γ be a finite set of PLTL formulas and σ a finite trace. The Γ-stutter factorization of σ is the unique increasing sequence of positions $\{i_0 \ldots i_k\}$ such that $i_0 = 0$, $i_k = |\sigma| - 1$ and:*

- *for each $\theta \in \Gamma$ and j, $0 \leq j < k$, the truth value of θ along $[i_j, i_{j+1})$ does not change, i.e., for all $h, l \in [i_j, i_{j+1})$, $(\sigma, h) \models \theta$ if and only if $(\sigma, l) \models \theta$.*
- *the truth value of some formula in Γ changes along adjacent segments (except at i_k), i.e., for each $0 \leq j < k - 1$, for some $\theta \in \Gamma$, $(\sigma, i_j) \models \theta$ if and only if $(\sigma, i_{j+1}) \not\models \theta$.*

In order to define the *relevant positions*, we define a finite version of the successor and predecessor relations. We use a special symbol $\mathbf{und}$ (meaning undefined) to denote the successor of the last point in the trace and the predecessor of the first point in the trace.

Definition 2 (Relativized successor). *Let Γ be a finite set of PLTL formulas and σ a finite trace. The relativized successor and relativized predecessor are defined as follows:*

$$succ_\Gamma(\sigma, i) := \begin{cases} \mathbf{und} & \text{If } i \geq |\sigma| - 1 \\ (\sigma, i_{j+1}) & \text{If } i \in [i_j, i_{j+1}) \text{ for some } j \in [0, k) \end{cases}$$

$$pred_\Gamma(\sigma, i) := \begin{cases} \mathbf{und} & \text{If } i = 0 \text{ or } i > |\sigma| - 1 \\ (\sigma, i_j) & \text{If } i \in (i_j, i_{j+1}] \text{ for some } j \in [0, k) \end{cases}$$

We extend the previous definitions to the trace assignment Π as follows; $succ_\Gamma(\mathbf{und}) = \mathbf{und}$ if there is a trace variable $x \in \text{VAR}$ such that the successor $succ_\Gamma(\Pi)(x) = \mathbf{und}$. Similarly, $pred_\Gamma(\Pi) = \mathbf{und}$ if there is a trace variable x such that the predecessor $pred_\Gamma(\Pi)(x) = \mathbf{und}$. Otherwise, $succ_\Gamma(\Pi)(x) = succ_\Gamma(\Pi(x))$ and $pred_\Gamma(\Pi)(x) = pred_\Gamma(\Pi(x))$ for each $x \in \text{VAR}$.

Finally, we define the semantics of SC-HyperLTL as follows:

$$\begin{aligned}
\Pi \models_T \exists x.\alpha &\Leftrightarrow \text{for some } \sigma \in T,\, \Pi[x \mapsto (\sigma,0)] \models_T \alpha \\
\Pi \models_T \forall x.\alpha &\Leftrightarrow \text{for all } \sigma \in T,\, \Pi[x \mapsto (\sigma,0)] \models_T \alpha \\
\Pi \models_T \Gamma.\varphi &\Leftrightarrow (\Pi,\Gamma) \models \varphi \\
(\Pi,\Gamma) \models a[x] &\Leftrightarrow a \in \sigma(p),\ \text{where } (\sigma,p) = \Pi(x) \\
(\Pi,\Gamma) \models \varphi_1 \,\mathbf{U}\, \varphi_2 &\Leftrightarrow \text{for some } j \geq 0\ \ succ^j_\Gamma(\Pi) \neq \mathbf{und},\, (succ^j_\Gamma(\Pi),\Gamma) \models \varphi_2, \\
&\qquad \text{and for all } i < j,\, (succ^i_\Gamma(\Pi),\Gamma) \models \varphi_1 \\
(\Pi,\Gamma) \models \mathbf{X}\varphi &\Leftrightarrow succ_\Gamma(\Pi) \neq \mathbf{und} \text{ and } (succ_\Gamma(\Pi),\Gamma) \models \varphi \\
(\Pi,\Gamma) \models \varphi_1 \,\mathbf{S}\, \varphi_2 &\Leftrightarrow \text{for some } j \geq 0\ \ pred^j_\Gamma(\Pi) \neq \mathbf{und},\, (pred^j_\Gamma(\Pi),\Gamma) \models \varphi_2, \\
&\qquad \text{and for all } i < j,\, (pred^i_\Gamma(\Pi),\Gamma) \models \varphi_1 \\
(\Pi,\Gamma) \models \mathbf{Y}\varphi &\Leftrightarrow pred_\Gamma(\Pi) \neq \mathbf{und} \text{ and } (pred_\Gamma(\Pi),\Gamma) \models \varphi \\
(\Pi,\Gamma) \models \langle x \rangle \beta[x] &\Leftrightarrow (\sigma,i) \models \beta \text{ with } (\sigma,i) = \Pi(x)
\end{aligned}$$

Theorem 1. *Model Checking of SC-HyperLTL is decidable.*

The proof proceeds by interpreting the logic on infinite traces, which is proven decidable in [8] via a reduction into QPTL (quantified propositional LTL) [36]. The resulting QPTL formula encodes the context and stuttering modalities using special padding letters and encoding the states of the model, and additionally requiring self-loop fresh states to create infinite traces from models with finite traces. The encoding from SC-HyperLTL to the logic defined in [8] can be found in the Appendix of the extended version of the manuscript [9].

Reversing Formulas. We now introduce an interesting property of our logic on finite traces. We first define the reverse function on formulas.

Definition 3 (Reverse SC-HyperLTL formula). *Let φ be a SC-HyperLTL formula. We define $rev(\varphi)$ as follows (here, $\Gamma_R := \{\mathbf{X}\,rev(\theta)|\theta \in \Gamma\}$):*

$$\begin{aligned}
rev(\forall x.\varphi) &:= \forall x.rev(\varphi) & rev(\mathbf{X}\psi) &:= \mathbf{Y}\,rev(\psi) \\
rev(\exists x.\varphi) &:= \exists x.rev(\varphi) & rev(\mathbf{Y}\psi) &:= \mathbf{X}\,rev(\psi) \\
rev(\Gamma.\psi) &:= \Gamma_R.rev(\psi) & rev(\langle x \rangle \psi) &:= \langle x \rangle\,rev(\psi) \\
rev(v[x]) &:= v[x] & rev(\psi_1 \,\mathbf{U}\, \psi_2) &:= rev(\psi_1) \,\mathbf{S}\, rev(\psi_2) \\
rev(\psi_1 \vee \psi_2) &:= rev(\psi_1) \vee rev(\psi_2) & rev(\psi_1 \,\mathbf{S}\, \psi_2) &:= rev(\psi_1) \,\mathbf{U}\, rev(\psi_2) \\
rev(\neg\psi) &:= \neg rev(\psi)
\end{aligned}$$

The reverse of a formula considers the trace in the opposite time direction, which is enabled by having future and past operators and traces having a beginning and an end. Since Γ points are evaluated considering the points with different past evaluation, we need to reverse Γ as well. Given a trace σ we use σ^{-1} for the trace such that $\sigma^{-1}(i) = \sigma(N - i)$ where $N = |\sigma| - 1$. We use Π^{-1} for the trace assignment that assigns σ^{-1} to x when Π assigns σ to x. The following holds directly from the definition.

Theorem 2. *For every formula φ, $rev(rev(\varphi)) \equiv \varphi$. Given trace assignment Π, $(\Pi,\Gamma) \models \varphi$ if and only if $(\Pi^{-1},\Gamma_R) \models rev(\varphi)$.*

Properties Expressible in SC-HyperLTL. We now show some important practical properties that can be expressed in the decidable fragment of the logic (others include observation determinism, initial-state opacity, etc.).

GMNI [26]: Goguen and Meseguer non-interference states that the observations of public information do not change when the secret information is removed. We can define this property for the asynchronous setting with the following formula:

$$\forall x.\exists y.\{lo\}.\langle x\rangle \mathbf{G}\lambda[x] \wedge \mathbf{G}(lo[x] \leftrightarrow lo[y]),$$

where *lo* represents the public information and λ represents the absence of secret inputs.

Observational Determinism [38]: Observational determinism states that if two executions have the same initial low input, then the two executions will be indistinguishable observing only low output. In the asynchronous setting we can express it with the following property, where *LI* and *LO* are respectively the sets of low inputs and low outputs.

$$\forall x.\forall y.\{LO\}. \bigwedge_{v \in LI} (v[x] \leftrightarrow v[y]) \rightarrow \bigwedge_{v \in LO} \mathbf{G}(v[x] \leftrightarrow v[y]),$$

Battery Sensor. Consider the high level model shown in Fig. 1, which describes a battery sensor. To deal with hardware faults in the sensor, a second sensor is used as a backup, and a Selector component chooses which sensor output (sensed1/sensed2) shall be considered. The selector component is connected with the Recovery component which detects faults like abnormal oscillations and triggers a recovery action by sending a "recovery" signal (dashed blue line). The Selector component will then choose Sensor2 as the main sensor. We assume that faults are permanent and that sensors run asynchronously.

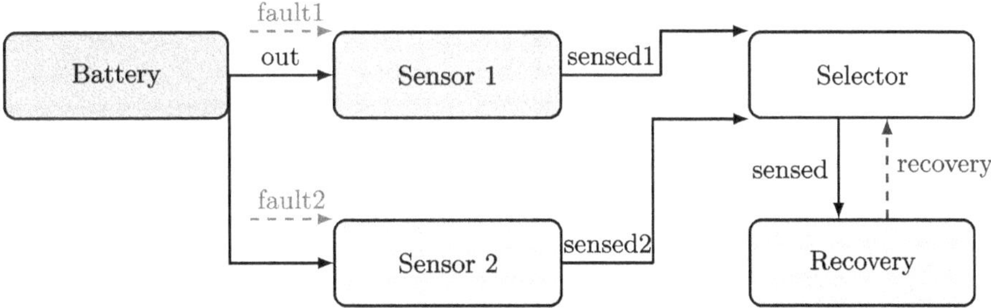

Fig. 1. Redundant Sensor System with Fault Detection, Isolation, and Recovery component schema. The coloured rectangles represents the internal components, the black arrow represent the data-port connection between components, the red dashed arrows represent fault events and, the blue dashed arrow represent the recovery signal. (Color figure online)

For any execution with a fault, all observationally equivalent executions (from the Recovery component perspective) trigger a recovery command within a bounded amount of steps, where an observation occurs when *sensed* changes:

$$\varphi_{rec} := \forall x.\forall y.\{sensed\}.\mathbf{G}(obseq \wedge \langle x \rangle faulted[x] \rightarrow \langle y \rangle \mathbf{F}^{\leq d} rcv[y])$$

where $\mathbf{F}^{\leq d}\phi$ is the bounded future operator defined as $\phi \vee \mathbf{X}(\mathbf{F}^{\leq d-1}\phi)$ with $\mathbf{F}^{\leq 0} := \phi$, $obseq := \mathbf{H}(sensed[x] \leftrightarrow sensed[y])$, $faulted := (\neg chd(sched) \mathbf{S} fault)$ and *chd(sched)* determines whether the sensed value changed.

The set $\{sensed\}$ is used to consider only the relevant points in which the sensed value changes. The formula *obseq* ensures that the two traces have the same sequences of sensed values. Then, *faulted*$[x]$ states that a fault occurred between the last observation point and the current one. Finally, the consequence is that the recovery command is activated within d steps in the related trace y.

4   Model Checking SC-HyperLTL

We now introduce two algorithms to verify SC-HyperLTL. These methods are defined for the finite-trace semantics studied in this paper, but they can be easily adapted to infinite-trace semantics.

4.1   Algorithm 1: Stuttering Encoding

The main idea is to extend the transition system to encode stuttering and introduce sub-formulas that encode alignment constraints for the observation points. The resulting transition system and formula can be verified using standard HyperLTL model checking. The model transformation is as follows.

Definition 4 (Stuttered extension). *Let M be a STS. We define the stuttered extension of M as $M^{st} := \langle V \cup \{st, halt\}, I \wedge \neg halt, T^{st} \rangle$ where $st, halt$ are two fresh Boolean variables and $T^{st} = ((st \vee halt) \rightarrow \bigwedge_{v \in V}(v' = v)) \wedge ((\neg st \wedge \neg halt) \rightarrow T) \wedge (halt \rightarrow halt')$ modifies T including stuttering transitions and halting self-loops.*

Essentially, M^{st} duplicates every state of M with a new version that models stuttering. M^{st} also adds a halting self-loop for every final state, so every accepting word of M is extended into an infinite word in M^{st}.

Definition 5 (Alignment under stuttering). *Let Γ be a finite set of PLTL formulae, $x_1, \ldots, x_n$ be a collection of trace variables. We define θ_{Γ}^{st} as follows:*

$$\theta_{\Gamma}^{st} := \mathbf{G}(\neg end^{st} \rightarrow \bigwedge_{1 < i \leq n}(\theta_{\Gamma}[x_i] \leftrightarrow \theta_{\Gamma}[x_1])), \qquad where$$

- $\theta_{\Gamma} := first^{st} \vee last^{st} \vee \bigvee_{\theta \in \Gamma}(\mathbf{Y}\theta \leftrightarrow \neg\theta)$ *represents the observation points,*
- $end^{st} := \bigvee_{1 \leq i \leq n} halt[x_i]$ *models the termination of some trace,*

- $\mathit{first}^{st} := \mathbf{Z}\bot$ *captures the first state of a trace (the same for all traces), and*
- $\mathit{last}^{st} := \neg \mathit{halt} \wedge \mathbf{X}\mathit{halt}$ *represents the last state.*

Definition 5 introduces the alignment constraints of the traces at the observation points. The alignment is preserved up to the "end" of the "shortest" stuttered trace. The formula θ_Γ captures observation points, which include the initial state first^{st} and the final state of the trace last^{st}. The antecedent of the implication holds whenever one of the traces has not surpassed the final state yet, while the consequent is the same truth value of observations of each trace. This constraint synchronously evaluates the observation points by padding the non-aligned observations using stuttering expansion.

We now introduce a variation of rewriting of formulas from [10] adapted to SC-HyperLTL. This rewriting $\mathcal{R}$ takes two quantifier-free and Γ-free SC-HyperLTL formulae ψ and φ_{st} and produces a quantifier-free HyperLTL formula:

Definition 6 (Stuttering Rewriting of SC-HyperLTL). *Let φ be a quantifier free SC-HyperLTL formula without Γ operator. We define the rewriting $\mathcal{R}_{\varphi_{st}}(\psi)$ as follows:*

$$\mathcal{R}_{\varphi_{st}}(v[x]) := v[x]$$
$$\mathcal{R}_{\varphi_{st}}(\neg\psi) := \neg\mathcal{R}_{\varphi_{st}}(\psi)$$
$$\mathcal{R}_{\varphi_{st}}(\psi_1 \vee \psi_2) := \mathcal{R}_{\varphi_{st}}(\psi_1) \vee \mathcal{R}_{\varphi_{st}}(\psi_2)$$
$$\mathcal{R}_{\varphi_{st}}(\langle x\rangle\psi) := \mathcal{R}_{st[x]\vee halt[x]}(\psi)$$
$$\mathcal{R}_{\varphi_{st}}(\mathbf{X}\psi) := \mathbf{X}(\varphi_{st}\,\mathbf{U}\,(\neg\varphi_{st}\wedge\mathcal{R}_{\varphi_{st}}(\psi)))$$
$$\mathcal{R}_{\varphi_{st}}(\mathbf{Y}\psi) := \mathbf{Y}(\varphi_{st}\,\mathbf{S}\,(\neg\varphi_{st}\wedge\mathcal{R}_{\varphi_{st}}(\psi)))$$
$$\mathcal{R}_{\varphi_{st}}(\psi_1\,\mathbf{U}\,\psi_2) := (\varphi_{st}\vee\mathcal{R}_{\varphi_{st}}(\psi_1))\,\mathbf{U}\,(\neg\varphi_{st}\wedge\mathcal{R}_{\varphi_{st}}(\psi_2))$$
$$\mathcal{R}_{\varphi_{st}}(\psi_1\,\mathbf{S}\,\psi_2) := (\varphi_{st}\vee\mathcal{R}_{\varphi_{st}}(\psi_1))\,\mathbf{S}\,(\neg\varphi_{st}\wedge\mathcal{R}_{\varphi_{st}}(\psi_2))$$

Finally, let $\varphi := \Gamma.\psi$ and let x_1 be a trace variables occuring in ψ. We denote $\mathcal{R}(\varphi) := \mathcal{R}_{\neg\theta_\Gamma[x_1]}.$

Temporal operator formulae are evaluated at positions where φ_{st} is false (that is, aligned at non-stuttering positions). For example, $\mathcal{R}_{\varphi_{st}}(\mathbf{X}p[x])$ would be true under the HyperLTL semantics only if at the next occurrence of $\neg\varphi_{st}$ the variable $p[x]$ is true. A detailed description of the PLTL part of the rewriting can be found in [10].

For singleton-context formulas, φ_{st} is replaced for the subformula with $st[x] \vee halt[x]$ that represents either a local stuttering state or the end of the local trace. The intuition is that, with singleton context, we want to only evaluate the points that are relevant for the local trace. For non-singleton formulae φ_{st} is set to $\neg\theta_\Gamma[x_1] \vee end^{st}$. Therefore, each temporal operator is evaluated only at the observation points of x_1 in which no trace has terminated yet. Provided the alignment of Definition 5, the observation points that need to be evaluated with finite semantics are the ones in which $\theta_\Gamma[x_1] \wedge \neg end^{st}$ is satisfied.

Theorem 3. *Let M be a STS and ψ be a quantifier-free SC-HyperLTL formula. Then,*

$$M \models \forall x_1 \ldots \forall x_n.\Gamma.\psi \quad \text{if and only if} \quad M^{st} \models \forall x_1 \ldots \forall x_n.\theta_\Gamma^{st} \to \mathcal{R}(\psi)$$
$$M \models \exists x_1 \ldots \exists x_n.\Gamma.\psi \quad \text{if and only if} \quad M^{st} \models \exists x_1 \ldots \exists x_n.\theta_\Gamma^{st} \wedge \mathcal{R}(\psi)$$

Proof. (Sketch) The constraint θ_Γ^{st} forces the traces to be aligned at the positions in which one of the elements of Γ changes or at the final or initial positions. Given this alignment, $\mathcal{R}$ parses Γ-temporal operators evaluating them only in the Γ positions of the first trace (which is aligned with the other traces). Note that in SC-HyperLTL context operators occur only with singleton contexts and with sub-formulae over the context trace, so $\mathcal{R}_{\varphi_{st}}(\langle x \rangle \psi)$ restricts the evaluation of ψ to the points in which $st[x] \vee halt[x]$ is false, that is, on the local points of the trace. It is easy to see that the original semantics is preserved by the translation. The complete proof is in the Appendix of the extended version of the manuscript [9].

4.2 Algorithm 2: Bounded Observations

We now introduce an alternative algorithm to deal with asynchrony for formulas with arbitrary quantifier alternations. This algorithm introduces k history variables $\{v_1, \ldots v_k\}$ for each variable v occurring inside the formula. The intended meaning of v_j is to mimic the value that v has at the j-th position of the stuttering factorization of the trace. Then, the variables are used in a BMC-like approach [7] to check whether the property holds on a horizon with at most k observations. If k observations are sufficient to prove or disprove the property then the outcome is informative. Note that the bound k is on the number of observations and not on the length of the trace, which is unbounded.

To simplify the encoding we consider formulae without singleton context expressions. Every single context sub-formula can be translated into automata with standard techniques, and the sub-formula can be replaced by a fresh variable as shown in Proposition 1 below. In this section, given an LTL_f formula β over variables V', we denote M_β as the STS with alphabet $V \cup \{v_\beta\}$ such that for each (finite) pointed trace (σ, i), $(\sigma, i) \models \beta \Leftrightarrow (\sigma, i) \models_f v_\beta$. We use $Sub(\varphi)$ for the set of all the sub-formulas of φ.

Proposition 1 (Removing singleton context). *Let M be a STS and α be a SC-HyperLTL formula. $M \models_f \alpha$ if and only if $(M \times \bigtimes_{\langle x_i \rangle \beta[x_i] \in Sub(\varphi)} M_\beta) \models_f \alpha'$, where α' is obtained by replacing each occurrence of $\langle x_i \rangle \beta[x_i]$ with $v_\beta[x]$.*

Definition 7 (Padding traces). *Let M be a STS, we define M^{pad} as the STS with a self-loop transition at the end of each execution: $M^{pad} := \langle V \cup \{end\}, I \wedge \neg end, (\neg end' \to T) \wedge (end' \to \bigwedge_{v \in V} v = v') \wedge (end \to end'), F \cup \{end\}\rangle$*

To deal with the fact that local traces might have different lengths, Def. 7 introduces the notion of padding traces by using the *end* variable.

Definition 8 (History-variable STS). *Let M be a STS, $\overline{V} \subseteq V$ be a set of variables, $\overline{V}^k := \{v_i \mid v \in \overline{V}, 0 \le i < k\}$ be the set containing k copies of the variables of $\overline{V}$ and pos be an enum variable ranging from 0 to $k + 1$.*

We define the STS $P^k_{\theta_\Gamma, \overline{V}} := \langle V \cup \overline{V}^k \cup \{pos, v_{\theta_\Gamma}, end\}, pos = 0, T^k \rangle$ with

$$T^k := \left(\begin{array}{c} (obs_\Gamma \rightarrow pos' = pos + 1) \wedge (\neg obs_\Gamma \rightarrow pos' = pos) \wedge \\ \bigwedge_{v_i \in \overline{V}^k} ((pos = i \wedge obs_\Gamma \rightarrow v'_i = v) \wedge (pos \ne i \vee \neg obs_\Gamma \rightarrow v'_i = v_i)) \end{array} \right)$$

where v_{θ_Γ} represents the points in which elements of Γ changed, $obs_\Gamma := \neg end \wedge (v_{\theta_\Gamma} \vee end' \vee pos = 0)$ represents the observation points in the trace i.e. either the points in which elements of Γ change, the initial state or the last point of the trace.

Definition 8 introduces the behaviour of history variables. Variable *pos* counts the observations seen so far, initialized to 0, and increased each time an observation is encountered. If, more than k observation are encountered *pos* takes the value $k + 1$ (encoding that more than k observations have happened). Each variable v_i mimics the value of v at the i-th observation. When $pos = i$, it takes the current value of v ($v'_i = v$), otherwise it stutters ($v'_i = v_i$). It should be noted that, before the i-th observation, v_i can take any value. Variable v_{θ_Γ} represents the points of observations, where one of the $\theta \in \Gamma$ changes. Figure 2 shows two traces of a system, with $\Gamma = \{a\}$ and $k = 3$.

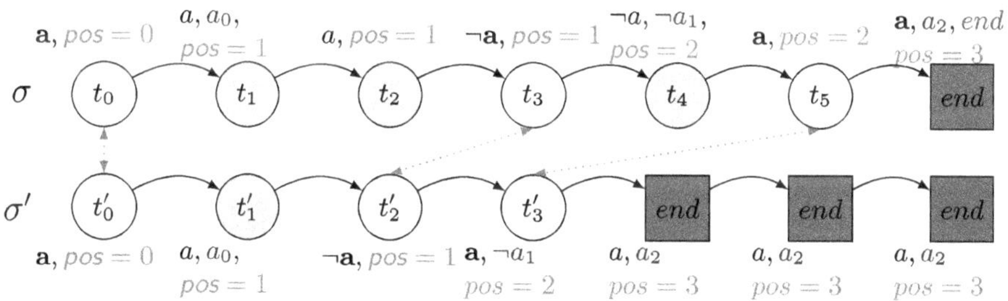

Fig. 2. Traces σ, σ' with the construction in Algorithm 2 for $k = 3$ and $\Gamma = \{a\}$. History variables are shown in blue; the end of the local traces are encoded by *end* variables, shown in purple; *pos* variables are shown in teal. Red dotted lines map the Γ-points of σ with the corresponding Γ-points of σ'. Pink circles represent the non-interesting trace points and purple boxes the end of the traces. The trace σ' is shorter than σ, so σ' is extended to be aligned with σ in the very last position according to Definition 7. (Color figure online)

Definition 9 (K-bound unrolling). *We define the bounded unrolling of quantifier free SC-HyperLTL formulas up to k. The complete encoding assumes formulae in negation normal form i.e. in which negations only occur over predicates.*

$${}^{\rho}[\![\varphi]\!]_k^k := \rho. \text{ For all } i < k:$$

$$
\begin{aligned}
{}^{\rho}[\![v[x]]\!]_i^k &:= v_i[x] & {}^{\rho}[\![\neg v[x]]\!]_i^k &:= \neg v_i[x] \\
{}^{\rho}[\![\psi_1 \vee \psi_2]\!]_i^k &:= {}^{\rho}[\![\psi_1]\!]_i^k \vee {}^{\rho}[\![\psi_2]\!]_i^k & {}^{\rho}[\![\psi_1 \wedge \psi_2]\!]_i^k &:= {}^{\rho}[\![\psi_1]\!]_i^k \wedge {}^{\rho}[\![\psi_2]\!]_i^k \\
{}^{\rho}[\![\mathbf{X}_\Gamma\psi]\!]_i^k &:= \neg end_\Gamma^{i+1} \wedge {}^{\rho}[\![\psi]\!]_{i+1}^k & {}^{\rho}[\![\mathbf{N}_\Gamma\psi]\!]_i^k &:= end_\Gamma^{i+1} \vee {}^{\rho}[\![\psi]\!]_{i+1}^k
\end{aligned}
$$

$$
{}^{\rho}[\![\psi_1 \,\mathbf{U}_\Gamma\, \psi_2]\!]_i^k := {}^{\rho}[\![\psi_2]\!]_i^k \vee ({}^{\rho}[\![\psi_1]\!]_i^k \wedge {}^{\rho}[\![\mathbf{X}\top]\!]_{i+1}^k \wedge {}^{\rho}[\![\psi_1 \,\mathbf{U}_\Gamma\, \psi_2]\!]_{i+1}^k)
$$

$$
{}^{\rho}[\![\psi_1 \,\mathbf{R}_\Gamma\, \psi_2]\!]_i^k := {}^{\rho}[\![\psi_2]\!]_i^k \wedge ({}^{\rho}[\![\psi_1]\!]_i^k \vee {}^{\rho}[\![\mathbf{N}\bot]\!]_{i+1}^k \vee {}^{\rho}[\![\psi_1 \,\mathbf{R}_\Gamma\, \psi_2]\!]_{i+1}^k)
$$

$$
{}^{\rho}[\![\mathbf{Y}_\Gamma\psi]\!]_0^k := \bot \quad {}^{\rho}[\![\mathbf{Z}_\Gamma\psi]\!]_0^k := \top \quad {}^{\rho}[\![\mathbf{Y}_\Gamma\psi]\!]_i^k := {}^{\rho}[\![\psi]\!]_{i-1}^k \quad {}^{\rho}[\![\mathbf{Z}_\Gamma\psi]\!]_i^k := {}^{\rho}[\![\psi]\!]_{i-1}^k
$$

$$
{}^{\rho}[\![\psi_1 \,\mathbf{S}_\Gamma\, \psi_2]\!]_i^k := {}^{\rho}[\![\psi_2]\!]_i^k \vee ({}^{\rho}[\![\psi_1]\!]_i^k \wedge {}^{\rho}[\![\mathbf{Y}\top]\!]_i^k \wedge {}^{\rho}[\![\psi_1 \,\mathbf{S}_\Gamma\, \psi_2]\!]_{i-1}^k)
$$

$$
{}^{\rho}[\![\psi_1 \,\mathbf{T}_\Gamma\, \psi_2]\!]_i^k := {}^{\rho}[\![\psi_2]\!]_i^k \wedge ({}^{\rho}[\![\psi_1]\!]_i^k \vee ({}^{\rho}[\![\mathbf{Z}\bot]\!]_i^k \vee {}^{\rho}[\![\psi_1 \,\mathbf{T}_\Gamma\, \psi_2]\!]_{i-1}^k))
$$

where $end_\Gamma^i := \bigvee_{x \in VAR}(pos[x] \leq i)$, $\rho \in \{\bot, \top\}$.

The formula ${}^{\rho}[\![\psi]\!]_0^k$ introduces an unrolling of formulas following the finite semantics of the operators, using the history variables as if they were synchronous projections of each trace and then traverse the traces synchronously. Since each trace might have a different number of observations, the encoding considers the end of the projected trace as the position i such that at least one trace has $pos = i$ (end_Γ^i).

The following theorem establishes the correctness of the reduction from SC-HyperLTL to HyperLTL with finite semantics using Definition 7, Definition 8 and, Definition 9.

Theorem 4. *Let M be a STS, let $\varphi := \mathcal{Q}_n x_n \ldots \mathcal{Q}_1 x_1.\Gamma.\psi$ be a SC-HyperLTL formula, where each $\mathcal{Q}_i \in \{\forall, \exists\}$ is a quantifier, let $\theta_\Gamma := \bigvee_{\theta \in \Gamma}(\neg\theta \leftrightarrow \mathbf{Y}\theta)$. Consider $M_\mathcal{B}^k := M^{pad} \times M_{\theta_\Gamma} \times P_{\theta_\Gamma,\overline{V}}^k$ to be the STS constructed by the synchronous composition of $P_{\theta_\Gamma,\overline{V}}^k$ with M and M_{θ_Γ} extended with end transitions. If $M_\mathcal{B}^k \models_f \mathbf{G}(pos \leq k)$ holds, then $M \models_f \mathcal{Q}_n x_n \ldots \mathcal{Q}_1 x_1.\psi$ iff $M_\mathcal{B}^k \models_f \mathcal{Q}_n x_n \ldots \mathcal{Q}_1 x_1.\mathbf{G}(\mathbf{N}\bot \to {}^{\bot}[\![\psi]\!]_0^k)$. Moreover,*

- *If $M_\mathcal{B}^k \models_f \mathcal{Q}_n x_n \ldots \mathcal{Q}_1 x_1.\mathbf{G}(\mathbf{N}\bot \to {}^{\bot}[\![\psi]\!]_0^k)$ then $M \models_f \mathcal{Q}_n x_n \ldots \mathcal{Q}_1 x_1.\psi$.*
- *If $M_\mathcal{B}^k \not\models_f \mathcal{Q}_n x_n \ldots \mathcal{Q}_1 x_1.\mathbf{G}(\mathbf{N}\bot \to {}^{\top}[\![\psi]\!]_0^k)$ then $M \not\models_f \mathcal{Q}_n x_n \ldots \mathcal{Q}_1 x_1.\psi$*

The theorem establishes a correspondence between the model-checking of a SC-HyperLTL formula in the original STS M and in the synchronous composition of M^{pad}, M_{θ_Γ} and $P_{\theta_\Gamma,\overline{V}}^k$. The extended system $M_\mathcal{B}^k$ incorporates the evaluation of Γ points using M_{θ_Γ}. The extended system considers traces with possibly different lengths via M^{pad}, and incorporates $P_{\theta_\Gamma,\overline{V}}^k$ to keep track of the assignments of variables in the observational point. The first part of the theorem states that, if M has at most k observations, then the unrolling ${}^{\rho}[\![\psi]\!]_0^k$ is satisfied at the end of the traces (i.e. when $\mathbf{N}\bot$ is true) if and only if the original model-checking problem holds. The second part of the theorem considers a partial relation between the model checking results of both problems. If the unrolling provides a positive result then the original formula satisfies the property as well. If proving the property requires more than k steps, the rule ${}^{\rho}[\![\psi]\!]_k^k$ returns $\bot$ as result (with $\rho = \bot$). Conversely, with an "optimistic" view by setting $\rho = \top$, a negative result (i.e. the property is not satisfied) entails that M violates the original formula. This mimics the pessimistic (resp. optimistic) semantics [30] for synchronous BMC with the difference that in our algorithm the actual traces have unbounded length.

Beyond k-Observations. It is possible to extend Algorithm 2 relaxing the global limit of k-observations into k-difference-in-observations, where observations are stored and evaluated modulo k. In Algorithm 2 the i-th observation is stored at point i (assuming $i < k$) and remains fixed for the rest of the trace. In the extended approach, the storage of observations would cycle (indexed by $i \mod k$), allowing properties to be checked dynamically over the trace. To support this, an additional constraint needs to keep track of which trace is the fastest, enabling evaluation relative to trace alignment. This modified algorithm can handle systems where the differences in observations is bounded by k.

5 Proof of Concept Implementation

We implemented the techniques described in Sect. 4 in a proof-of-concept implementation using the nuXmv model checker [16]. The implementation applies the translations described above and outputs either a self-composed SMV model (if the property has no quantifier alternation) or an AutoHyper model (see [6]) with its corresponding HyperLTL property.

To validate our approach, we use three examples: (1) the example in Sect. 1, (2) a variation of the battery sensor from Sect. 3 and (3) the parallel composition, under interleaving semantics, of two programs: *P0*, that can take an arbitrary number of steps to terminate, and *P1*, which terminates after k steps. A scheduler that decides non-deterministically which process runs: *P0* (if $sched = 0$) or *P1* (if $sched = 1$). We intend to prove that the execution of *P1* is *deterministic* no matter what *P0* does, encoded as follows:

$$\psi_{det} := \forall x. \forall y. \{P1.var\}. \langle x \rangle FSchd[x] \wedge \langle y \rangle FSchd[y] \rightarrow \mathbf{G}(P1.var[x] \leftrightarrow P1.var[y])$$

where $FSchd := \mathbf{G}(\mathbf{NN}\bot \rightarrow sched = 1)$ states that the last scheduled process is *P1*. The property states that, if in the last transition of both traces *P1* is scheduled, then the two programs have always the same value of variable *P1.var*. We analyzed variations with quantifier alternations, incorrect versions (with manual bugs), and versions with unbounded amount of observations and controlled scheduling for bounded observations. The experiments[1] were run on a cluster with Intel Xeon CPU 6226R nodes running at 2.9 GHz with 32CPU, 12 GB. The timeout for each run was one hour and the memory cap was set to 1 GB with 1 CPU assigned for each instance. Table 1 reports the execution times of the approaches. We observe that nuXmv is able to solve quite rapidly all instances, but AutoHyper suffers a significant performance slowdown. One possible reason is that encoding the behavior as part of the formula has a significant impact on

---

[1] All the data can be found at https://es-static.fbk.eu/people/bombardelli/papers/ spin25/spin25.tar.gz.

explicit state model-checkers like AutoHyper. Another possibility is the additional variables introduced to deal with traces of different lengths. We leave the research of how to alleviate this problem as future work. More in general, we observe that Algorithm 1, that applies stuttering rewriting (for quantifier-alternation free instances) is more successful (when applicable).

Table 1. Empirical evaluation, where $*$ means non-informative with k-bound due to too many observations (TO is 1 h). Dashed line on tools mean that they cannot be solved with it. For instance q.alt. model cannot be checked with nuXmv and models with past (Battery Sensor) cannot be checked with AutoHyper.

Name	Method	Bound	Outcome	nuXmv Time (s)	AutoHyper Time (s)
Process $n = 2$	k-bound (Alg.2)	3	True	2.82	5.54
Process $n = 4$	k-bound (Alg.2)	5	True	4.03	TO
Process $n = 4$	Stuttering(Alg.1)	–	True	1.86	MO
Process $n = 6$	k-bound (Alg.2)	7	True	4.17	MO
Process $n = 6$	Stuttering(Alg.1)	–	True	2.65	MO
Process $n = 6$ (bug)	k-bound (Alg.2)	7	False	3.33	MO
Process $n = 6$ (bug)	Stuttering(Alg.1)	–	False	3.02	MO
Process q. alt. $n = 2$	k-bound (Alg.2)	3	True	–	34.93
Process q. alt. $n = 3$	k-bound (Alg.2)	4	True	–	TO
Motivating example	k-bound (Alg.2)	7	True	1245.95	TO
Motivating example	k-bound (Alg.2)	3	False*	34.64	TO
Motivating example	Stuttering(Alg.1)	–	True	20.88	6.19
Battery Sensor	k-bound (Alg.2)	7	False*	11.86	–
Battery Sensor	Stuttering(Alg.1)	–	True	6.03	–
Battery Sensor (bug)	Stuttering(Alg.1)	–	True	3.04	–

6 Conclusion

We introduced SC-HyperLTL, a novel temporal logic with past for expressing hyperproperties over finite traces, that combines features from two existing asynchronous hyperlogics: stuttering HyperLTL and context HyperLTL. SC-HyperLTL addresses the challenge of verifying systems with finite traces, where events occur asynchronously and past operators are required. We established the decidability of model-checking SC-HyperLTL and proposed two practical algorithms for its verification: one based on stuttering translation and the other that leverages auxiliary variables. We implement these in a proof-of-concept prototype based on the nuXmv model checker, showing the feasibility of the approach.

One major direction for future work is the verification of decidable extensions of our logic [8], which currently lacks practical verification techniques. Another future direction is to improve the performance. For Algorithm 1, one possible direction is to try to reduce the problem to invariant checking. For Algorithm 2, we will study how to relax the k-observation assumption. Finally, although the decidability proof provides an implicit upper bound on complexity, it would be interesting to determine the exact complexity of our logics. Additionally, it would be valuable to precisely characterize the complexity of both the stuttering and k-bound algorithms.

Acknowledgements. This Work was funded in part by the DECO Project (PID2022-138072OB-I00) funded by MCIN/AEI/10.13039/501100011033 and by the ESF+.

References

1. E. Bartocci, T. Ferrére, T. A. Henzinger, D. Nickovic, and A. O. da Costa. Flavors of sequential information flow. In *Proc. of the 23rd Int'l Conf. on Verification, Model Checking, and Abstract Interpretation (VMCAI'22)*, 2022
2. E. Bartocci, T. A. Henzinger, D. Nickovic, and A. Oliveira da Costa. Hypernode Automata. In G. A. Pérez and J.-F. Raskin, editors, *34th International Conference on Concurrency Theory (CONCUR 2023)*, volume 279 of *Leibniz International Proceedings in Informatics (LIPIcs)*, pages 21:1–21:16, Dagstuhl, Germany, 2023. Schloss Dagstuhl – Leibniz-Zentrum für Informatik
3. Baumeister, J., Coenen, N., Bonakdarpour, B., Finkbeiner, B., Sánchez, C.: A Temporal Logic for Asynchronous Hyperproperties. In: Silva, A., Leino, K.R.M. (eds.) CAV 2021. LNCS, vol. 12759, pp. 694–717. Springer, Cham (2021). https://doi.org/10.1007/978-3-030-81685-8_33
4. Beutner, R.: Automated software verification of hyperliveness. In: Finkbeiner, B., Kovács, L. (eds.) Tools and Algorithms for the Construction and Analysis of Systems. pp, pp. 196–216. Springer Nature Switzerland, Cham (2024)
5. Beutner, R., Finkbeiner, B.: Software verification of hyperproperties beyond k-safety. In: Shoham, S., Vizel, Y. (eds.) Computer Aided Verification. pp, pp. 341–362. Springer International Publishing, Cham (2022)
6. R. Beutner and B. Finkbeiner. AutoHyper: Explicit-state model checking for HyperLTL. In *Proc. of the 29th Int'l Conf. on Tools and Algorithms for the Construction and Analysis of Systems (TACAS'23)*, volume 13993 of *LNCS*, pages 145–163. Springer, 2023
7. A. Biere, A. Cimatti, E. M. Clarke, and Y. Zhu. Symbolic model checking without BDDs. In *Proc. of the 5th Int'l Conf. on Tools and Algorithms for Construction and Analysis of Systems (TACAS'99)*, volume 1579 of *LNCS*, pages 193–207. Springer, 1999
8. A. Bombardelli, L. Bozzelli, C. Sánchez, and S. Tonetta. Unifying asynchronous logics for hyperproperties. *CoRR*, abs/2404.16778, 2024
9. A. Bombardelli, L. Bozzelli, C. Sánchez, and S. Tonetta. (asynchronous) temporal logics for hyperproperties on finite traces - extended version. 2025. https://es-static.fbk.eu/people/bombardelli/papers/spin25/spin25_ext.pdf
10. A. Bombardelli and S. Tonetta. Asynchronous composition of local interface LTL properties. In *Proc. of the 14th Int'l NASA Formal Methods Symposium (NFM'22)*, volume 13260 of *LNCS*, pages 508–526, 2022

11. Bonakdarpour, B., Prabhakar, P., Sánchez, C.: Model Checking Timed Hyperproperties in Discrete-Time Systems. In: Lee, R., Jha, S., Mavridou, A., Giannakopoulou, D. (eds.) NFM 2020. LNCS, vol. 12229, pp. 311–328. Springer, Cham (2020). https://doi.org/10.1007/978-3-030-55754-6_18

12. Bonakdarpour, B., Sanchez, C., Schneider, G.: Monitoring Hyperproperties by Combining Static Analysis and Runtime Verification. In: Margaria, T., Steffen, B. (eds.) ISoLA 2018. LNCS, vol. 11245, pp. 8–27. Springer, Cham (2018). https://doi.org/10.1007/978-3-030-03421-4_2

13. M. Bozzano, A. Cimatti, M. Gario, and S. Tonetta. Formal design of asynchronous fault detection and identification components using temporal epistemic logic. *Logical Methods in Computer Science*, Volume 11, Issue 4, Nov. 2015

14. L. Bozzelli, A. Peron, and C. Sánchez. Asynchronous extensions of HyperLTL. In *Proc. of the 36th Annual ACM/IEEE Symposium on Logic in Computer Science (LICS'21)*, pages 1–13. IEEE, 2021

15. L. Bozzelli, A. Peron, and C. Sánchez. Expressiveness and decidability of temporal logics for asynchronous hyperproperties. In *Proc. of the 33rd International Conference on Concurrency Theory (CONCUR'22)*, volume 243 of *Leibniz International Proceedings in Informatics (LIPIcs)*, pages 27:1–27:16, Dagstuhl, Germany, 2022. Schloss Dagstuhl – Leibniz-Zentrum für Informatik

16. R. Cavada, A. Cimatti, M. Dorigatti, A. Griggio, A. Mariotti, A. Micheli, S. Mover, M. Roveri, and S. Tonetta. The nuxmv symbolic model checker. In *International Conference on Computer Aided Verification*, 2014

17. Clarkson, M.R., Finkbeiner, B., Koleini, M., Micinski, K.K., Rabe, M.N., Sánchez, C.: Temporal Logics for Hyperproperties. In: Abadi, M., Kremer, S. (eds.) POST 2014. LNCS, vol. 8414, pp. 265–284. Springer, Heidelberg (2014). https://doi.org/10.1007/978-3-642-54792-8_15

18. Clarkson, M.R., Schneider, F.B.: Hyperproperties. J. Comput. Secur. **18**(6), 1157–1210 (2010)

19. N. Coenen, B. Finkbeiner, C. Hahn, and J. Hofmann. The hierarchy of hyperlogics. In *Proc. of the 34th Annual ACM/IEEE Symposium on Logic in Computer Science (LICS'19)*, pages 1–13. IEEE, 2019

20. Coenen, N., Finkbeiner, B., Sánchez, C., Tentrup, L.: Verifying Hyperliveness. In: Dillig, I., Tasiran, S. (eds.) CAV 2019. LNCS, vol. 11561, pp. 121–139. Springer, Cham (2019). https://doi.org/10.1007/978-3-030-25540-4_7

21. A. Correnson, T. Nießen, B. Finkbeiner, and G. Weissenbacher. Finding ∀∃ Hyperbugs using Symbolic Execution. *Proc. ACM Program. Lang.*, 8(OOPSLA2), Oct. 2024

22. Eisner, C., Fisman, D., Havlicek, J., Lustig, Y., McIsaac, A., Van Campenhout, D.: Reasoning with Temporal Logic on Truncated Paths. In: Hunt, W.A., Somenzi, F. (eds.) CAV 2003. LNCS, vol. 2725, pp. 27–39. Springer, Heidelberg (2003). https://doi.org/10.1007/978-3-540-45069-6_3

23. B. Finkbeiner, M. Rabe, and C. Sánchez. Algorithms for model checking HyperLTL and HyperCTL*. In *In Proc. of the 27th Int'l Conf. on Computer Aided Verification (CAV'15)*, volume 9206 of *LNCS*, pages 30–48. Springer, 2015

24. G. D. Giacomo, P. Felli, M. Montali, and G. Perelli. HyperLDLf: a logic for checking properties of finite traces process logs. In *Proc. of the 30th Int'l Joint Conf. on Artificial Intelligence (IJCAI'21)*, pages 1859–1865. ijcai.org, 2021

25. G. D. Giacomo and M. Y. Vardi. Linear temporal logic and linear dynamic logic on finite traces. In *Proc. of the 23rd Int'l Joint Conference on Artificial Intelligence (IJCAI'13)*, pages 854–860. IJCAI/AAAI, 2013

26. J. A. Goguen and J. Meseguer. Security policies and security models. In *Proc. of the IEEE Symposium on Security and Privacy (S&P'82)*, pages 11–20. IEEE, 1982
27. J. O. Gutsfeld, M. Müller-Olm, and C. Ohrem. Propositional dynamic logic for hyperproperties. In *Proc. of the 31st Int'l Conf. on Concurrency Theory (CONCUR'20)*, LIPIcs 171, pages 50:1–50:22. Schloss Dagstuhl - Leibniz-Zentrum für Informatik, 2020
28. J. O. Gutsfeld, M. Müller-Olm, and C. Ohrem. Automata and fixpoints for asynchronous hyperproperties. *Proc. ACM Program. Lang.*, 5(POPL):1–29, 2021
29. Hsu, T.-H., Bonakdarpour, B., Finkbeiner, B., Sánchez, C.: Bounded model checking for asynchronous hyperproperties. In: Sankaranarayanan, S., Sharygina, N. (eds.) Tools and Algorithms for the Construction and Analysis of Systems, pp. 29–46. Springer Nature Switzerland, Cham (2023)
30. T.-H. Hsu, C. Sánchez, and B. Bonakdarpour. Bounded model checking for hyperproperties. In *Proc. of the 27th Int'l Conf on Tools and Algorithms for the Construction and Analysis of Systems (TACAS'21). Part I*, volume 12651 of *LNCS*, pages 94–112. Springer, 2021
31. Lichtenstein, O., Pnueli, A., Zuck, L.: The glory of the past. In: Parikh, R. (ed.) Logic of Programs 1985. LNCS, vol. 193, pp. 196–218. Springer, Heidelberg (1985). https://doi.org/10.1007/3-540-15648-8_16
32. Z. Manna and A. Pnueli. *The Temporal Logic of Reactive and Concurrent Systems*. Springer-Verlag, 1992
33. Z. Manna and A. Pnueli. *Temporal verification of reactive systems - safety*. Springer, 1995
34. A. Pnueli. The temporal logic of programs. In *Proc. of the 18th IEEE Symp. on Foundations of Computer Science (FOCS'77)*, pages 46–67. IEEE CS Press, 1977
35. M. N. Rabe. *A temporal logic approach to information-flow control*. PhD thesis, Saarland University, 2016
36. Sistla, A.P., Vardi, M.Y., Wolper, P.: The Complementation Problem for Büchi Automata with Applications to Temporal Logic. Theoret. Comput. Sci. **49**, 217–237 (1987)
37. Y. Wang, M. Zarei, B. Bonakdarpour, and M. Pajic. Statistical verification of hyperproperties for cyber-physical systems. *ACM Transactions on Embedded Computing systems (TECS)*, 18(5s):92:1–92:23, 2019
38. S. Zdancewic and A. C. Myers. Observational determinism for concurrent program security. In *Proc. of the 16th IEEE Computer Security Foundations Workshop (CSFW'03)*, pages 29–43. IEEE, 2003

An Efficient and Versatile Approach
to Shortest Path Problems
in Interprocedural Programs

Theo De Castro Pinto[1,2], Antoine Rollet[1(✉)], and Grégoire Sutre[1]

[1] Univ. Bordeaux, CNRS, Bordeaux INP, LaBRI, UMR 5800, 33400 Talence, France
{theo.de-castro-pinto,antoine.rollet,gregoire.sutre}@labri.fr
[2] Serma Safety and Security, 33600 Pessac, France

Abstract. This paper addresses the shortest path problem for weighted interprocedural automata (WIPA), a simple model adapted to the analysis of low-level programs. We propose an efficient and versatile approach relying on Knuth's generalization of Dijkstra's algorithm to context-free grammars. Several variants of the problem are considered in the paper, depending on the starting and ending stacks. Moreover, we solve both the single-source and single-target versions. Each problem is translated into a succint context-free grammar, permitting the efficient computation of a solution in $O(n \log n + m)$ time, where n and m are the number of states and transitions in the underlying WIPA. We provide experimental results on real-size programs showing the scalability of the approach.

Keywords: Interprocedural analysis · Shortest path problems · Context-free grammars · Dijkstra's algorithm · Program analysis · Static analysis

1 Introduction

Context. The shortest path problem is an important problem with applications in many areas, including interprocedural program analysis. Indeed, shortest distances can be used to guide symbolic execution [1,5,13], model checking [7,18] and fuzzing [3]. Dijkstra's algorithm provides an efficient solution to the shortest path problem on finite graphs [6], but the extension of this algorithm to interprocedural graphs is non-trivial. Both the structures and the algorithms used to compute shortest paths in interprocedural programs vary from paper to paper [1,3,15]. This leads to varying time and space complexities for these computations.

Contributions. In this paper we tackle the shortest path problem for weighted interprocedural automata (WIPA), a simple and flexible model for low-level (assembly or binary) programs with function calls. We address both the single-source and single-target versions of the problem. Informally, it consists in finding

G. Ernst and K. Y. Rozier (Eds.): SPIN 2025, LNCS 15945, pp. 44–65, 2026.
https://doi.org/10.1007/978-3-032-06847-7_3

the (shortest) distance from a source location to (resp., to a target location from) all other locations in the program. Our approach takes advantage of Knuth's generalization of Dijkstra's algorithm to context-free grammars. This generalization, presented in [11], can be used to efficiently compute for each nonterminal the minimal weight of a word in its language. In this paper:

- We propose an efficient translation into context-free grammars of the considered shortest path problems on WIPA. The translation is designed so that Knuth's algorithm [11] can be used to solve the shortest path problem.
- We define the shortest path problem with four levels of increasing precision with regard to the starting stack. We consider *arbitrary* stacks, *return edges*, *coherent* stacks, and *reachable* stacks. Our solution is more precise than the ones presented in [1] and [3], which are limited to return edges. Moreover, our solution has a time complexity of $O(n \log n + m)$, where n and m are the number of states and transitions in the underlying WIPA. This improves the complexities given in [1] and [3], which are respectively $O(f^2 m \log n)$ and $O(f^2 + fn^2)$, where f is the number of procedures in the program.
- We provide experimental results showing the efficiency and benefits of our approach. The SV-COMP benchmarks [2] are used for the experimentation.

Related Work. Graph reachability in the context of interprocedural programs is a well-studied problem, as it has various applications. Notably, Reps et al. introduced context-free language reachability problems in [16]. The key idea is to reduce dataflow analysis problems to graph reachability problems, where a path is a valid solution if its label respects some context-free grammar. Different grammars can be used such as realizable grammars [16], Dyck grammars [4,10,14] or PN grammars [8,15]. In [12,17], the authors advocate the use of weighted pushdown systems (WPDS) for interprocedural dataflow analysis. This approach is more general than the shortest path problem that we address in this paper, as the weight domain can be any idempotent semiring. But the time complexity suffers from this generality as the proposed algorithms run in $O(n^2 m)$ (resp., $O(n^3 m^2)$) time for bounded (resp., unbounded) semirings, with n and m the numbers of states and rules in the WPDS. Ma et al. propose in [13] to guide symbolic execution using shortest distances defined by way of a PN grammar. A similar grammar is used in [1] to solve the problem on visibly pushdown automata. In these works, grammars are used to define which paths are (interprocedurally) valid, whereas we use grammars to generate valid paths directly. In both [1] and [3], the time complexity depends on the number of procedures within the program. Our approach gets rid of this factor, and yields a time complexity akin to Dijkstra's algorithm. Moreover, it is adapted for low-level programs and more precise than existing approaches, such as [1], which are based on return edges.

2   Running Example

We illustrate the approach advocated in this paper on the running example depicted in Fig. 1. This example is what we call a *weighted interprocedural*

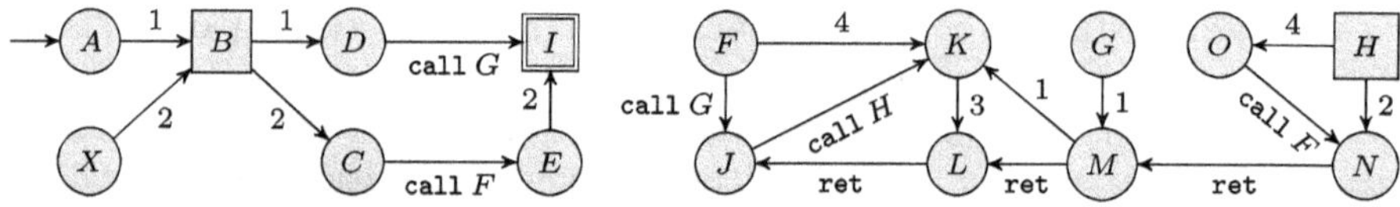

Fig. 1. A weighted interprocedural automaton. The source states are B and H, the target state is I and A is considered to be the initial state. Function calls and returns have an implicit weight of zero in this example. (Color figure online)

Table 1. A table showing the summary of each function along the second line, the (shortest) distance from the sources B and H to all states along the third line, and the (shortest) distance from all states to the target I in the last line.

	A	B	C	D	E	F	G	H	I	J	K	L	M	N	O	X
Summary						6	1	2								
S → All	+∞	0	2	1	5	2	1	0	2	3	2	5	2	2	4	+∞
All → T	3	2	8	1	2	8	1	7	0	7	3	0	0	5	11	+∞

automaton, a simple and flexible model, defined formally in Sect. 4, that is particularly convenient to abstract any kind of program with function calls, including low-level (assembly or binary) programs. This automaton comprises 16 nodes, that are called *states*, and 19 edges, that are called *transitions*. The *initial* state (i.e., the entry point of the program) is state A. Each transition is labelled by an *operation*, which is a function call, a function return, or an internal operation, and a *weight*, which is a nonnegative rational number. Functions are simply states in our setting (i.e., functions are identified with their entry point). The semantics of function calls and function returns are as expected. For instance, the function F is called on the transition from state C to state E. This means that from state C, control moves to state F, continues from there until it reaches a return transition, here $L \xrightarrow{0,\mathrm{ret}} J$, and then returns to state E. A run of the automaton is a sequence of steps between configurations, the latter being composed of a state and a stack. As expected, the stack keeps track of the return sites of active function calls. For instance, the configuration (G, JE) corresponds to state G with stack $J \cdot E$, shortly written JE. Here is a run showing how the stack behaves: $(C,\varepsilon) \xrightarrow{0} (F, E) \xrightarrow{0} (G, JE) \xrightarrow{1} (M, JE) \xrightarrow{0} (J, E)$. The weight of a run is the sum of the weights along the run, in the example above it is 1.

Given a weighted interprocedural automaton such as the one depicted in Fig. 1, various shortest path problems are of interest. One such problem is the computation of the summary of each function. By *summary* of a function, we mean the (shortest) distance, that is the minimal weight of a run, from the function's entry state, with empty stack, to an exit state[1] (here L, M or N), again with empty stack. For instance, the summary of function F is 6 and a witnessing run is $(F, \varepsilon) \overset{0}{\hookrightarrow} (G, J) \overset{1}{\hookrightarrow} (M, J) \overset{0}{\hookrightarrow} (J, \varepsilon) \overset{0}{\hookrightarrow} (H, K) \overset{2}{\hookrightarrow} (N, K) \overset{0}{\hookrightarrow} (K, \varepsilon) \overset{3}{\hookrightarrow} (L, \varepsilon)$. Function summaries for our running example can be found in the line *Summary* of Table 1. Note that summaries only apply to functions, so a state that is not the start of a function has no summary.

Another problem that can be tackled is the computation of the distance from a set of source states to all states. In our example (see Fig. 1), both B and H are source states (drawn with squares), so we want to determine for every state q the distance from $\{B, H\}$ to q. For instance, the distance from $\{B, H\}$ to E is 2 and a witnessing run is $(H, E) \overset{2}{\hookrightarrow} (N, E) \overset{0}{\hookrightarrow} (E, \varepsilon)$. However, this run may be considered as spurious because the stack E is not reachable in state H. Indeed, assuming that we start in the initial state A with an empty stack, one can show that the set of reachable stacks is $\{\varepsilon\}$ for state B and $\{KN\}^* \cdot \{KE\}$ for state H. So a more interesting problem is to determine for every state q the distance from $\{B, H\}$, with a reachable stack, to q. The resulting distance to E is 5 and a witnessing run is $(H, KE) \overset{2}{\hookrightarrow} (N, KE) \overset{0}{\hookrightarrow} (K, E) \overset{3}{\hookrightarrow} (L, E) \overset{\varepsilon}{\hookrightarrow} (E, \varepsilon)$. The complete solution for this problem is given in the line $S \to All$ of Table 1. The value of a state is $+\infty$ when there is no run to this state from a source state with a reachable stack.

The converse problem is of even higher interest due to its applications in symbolic execution [1,5,13], model checking [7,18] and fuzzing [3]. This problem is the computation of the distance from all states to a set of target states. Here, the only target state is I (drawn with a double square), so we want to determine for every state q the distance from q to the target state I. We may start in q with an arbitrary stack. But, as in the previous problem, it makes sense to only consider runs that start in q with a reachable stack. For instance, the distance from K to I is 3 and a witnessing run is $(K, I) \overset{3}{\hookrightarrow} (L, I) \overset{0}{\hookrightarrow} (I, \varepsilon)$. Clearly, I is a reachable stack for state K, since it can be generated by the run $(A, \varepsilon) \overset{*}{\hookrightarrow} (D, \varepsilon) \overset{0}{\hookrightarrow} (G, I) \overset{*}{\hookrightarrow} (K, I)$. The complete solution for this problem is given in the line $All \to T$ of Table 1. The value of a state is $+\infty$ when there is no run from this state, with a reachable stack, to the target state. In particular, the value for state X is $+\infty$. This comes from the fact that X is not reachable from the initial state A (whatever the stack).

In this paper, we present an efficient and flexible approach to solve all shortest path problems discussed in this section. Our approach is divided in two steps. First we transform the automaton into a fitting context-free grammar. Second, we use Knuth's generalization of Dijkstra's algorithm on the generated grammar. We recall the latter in the next section.

---

[1] In fact, we also account for the weight of the return transition in the function summary, see Remark 9. Return transitions have weight zero in our running example.

3   Knuth's Generalization of Dijkstra's Algorithm to Context-Free Grammars

In this section, we recall Knuth's generalization of Dijkstra's algorithm [6] to context-free grammars, presented in [11]. But before that, some basic definitions and notations on context-free grammars are needed.

3.1   Context-Free Grammars

A *context-free grammar*, or *CFG* for short, is a triple $\mathcal{G} = (V, \Sigma, P)$, where V and Σ are disjoint finite sets of *nonterminal* and *terminal* symbols, respectively, and $P \subseteq V \times (V \cup \Sigma)^*$ is a finite set of *productions*. Following the usual convention, a production (X, α) is also written $X \to \alpha$. We let $S = V \cup \Sigma$ denote the set of all *symbols*. A word $z \in S^*$ is called a *sentence*. We let $\Rightarrow$ denote the *derivation step* binary relation on sentences, defined as the set of pairs (z, z') in $S^* \times S^*$ such that there exist two sentences u, v in S^* and a production $X \to \alpha$ in P verifying $z = uXv$ and $z' = u\alpha v$. The reflexive and transitive closure of $\Rightarrow$ is written $\overset{*}{\Rightarrow}$. The *language* of a sentence z in S^* is the subset $L_z^{\mathcal{G}} \subseteq \Sigma^*$ defined by $L_z^{\mathcal{G}} = \{w \in \Sigma^* \mid z \overset{*}{\Rightarrow} w\}$.

3.2   Minimal Weights in Context-Free Grammars

We let $\mathbb{Q}_{\geq 0}$ and $\overline{\mathbb{R}}$ denote the sets of nonnegative rational numbers and of extended real numbers, respectively. Recall that $(\overline{\mathbb{R}}, \leq)$ is a complete lattice. The greatest lower bound of a set $X \subseteq \overline{\mathbb{R}}$ is written $\inf X$. Given a finite set Σ, a *weight function* for Σ is any function $\lambda : \Sigma \to \mathbb{Q}_{\geq 0}$. Such a weight function λ induces a monoid morphism from $(\Sigma^*, \cdot)$ to $(\mathbb{Q}_{\geq 0}, +)$, that we also denote by λ. So we have $\lambda(a_1 \cdots a_n) = \lambda(a_1) + \cdots + \lambda(a_n)$ for every word $a_1 \cdots a_n$ in Σ^*. In particular, $\lambda(\varepsilon) = 0$. For every subset L of Σ^*, we denote by $\lambda(L)$ the set $\{\lambda(w) \mid w \in L\}$. Observe that $\inf \lambda(L) = +\infty$ if L is empty, and $\inf \lambda(L) = \min \lambda(L) \in \mathbb{Q}_{\geq 0}$ otherwise. This observation easily follows from Higman's Lemma and the fact that we only allow nonnegative weights.

Now consider a context-free grammar $\mathcal{G} = (V, \Sigma, P)$ equipped with a weight function $\lambda : \Sigma \to \mathbb{Q}_{\geq 0}$. We introduce the *minimal weight* function $\mu : V \to (\mathbb{Q}_{\geq 0} \cup \{+\infty\})$ defined by $\mu(X) = \inf \lambda(L_X^{\mathcal{G}})$. The remainder of this section addresses the computation of the minimal weight function μ.

In the late 1970s, Knuth proposed in [11] a generalization to CFGs of Dijkstra's shortest path algorithm [6]. This generalization can be used to compute the minimal weight function μ defined above, but it can also be applied to other problems on CFGs. We provide in Algorithm 1 a detailed pseudocode of Knuth's algorithm, because the latter is only sketched in [11]. As explained in the comment at line 1, the algorithm starts with the construction of the arrays *cnt* and *dep*, whose sole purpose is to efficiently implement the enumeration at lines 9–14. The *tentative* minimal weights computed so far are maintained in the array ν. A

Algorithm 1. Knuth's Generalization of Dijkstra's Algorithm

Input: A CFG $\mathcal{G} = (V, \Sigma, P)$ and a weight function $\lambda : \Sigma \to \mathbb{Q}_{\geq 0}$.
Output: The minimal weight function $\mu : V \to (\mathbb{Q}_{\geq 0} \cup \{+\infty\})$.

```
 1: ▷ The algorithm starts by setting up two arrays named cnt and dep. The array cnt
       is indexed by P and the array dep is indexed by (V ∪ {⊥}), where ⊥ ∉ V is a new
       symbol. After this line, it holds that cnt[Y → α] = |{X ∈ V | X occurs in α}|
       for all (Y → α) ∈ P, that dep[⊥] = {(Y → α) ∈ P | α ∈ Σ*}, and that
       dep[X] = {(Y → α) ∈ P | X occurs in α} for all X ∈ V.                        ◁
 2: pq ← createPriorityQueue()
 3: foreach X ∈ V do
 4:     ν[X] ← +∞
 5:     pq.insert(X, +∞)
 6: X ← ⊥
 7: while not pq.isEmpty() do
 8:     ▷ cnt[Y → α] = |{X ∈ pq.items() | X occurs in α}| for all (Y → α) ∈ P      ◁
 9:     foreach (Y → α) ∈ dep[X] do
10:         if cnt[Y → α] = 0 then                        ▷ α ∈ ((V \ pq.items()) ∪ Σ)*
11:             m ← Σⁿᵢ₌₁(ν ⊎ λ)(αᵢ)              ▷ α = α₁ ⋯ αₙ with αᵢ ∈ (V ∪ Σ)
12:             if m < ν[Y] then
13:                 ν[Y] ← m
14:                 pq.decreaseKey(Y, m)
15:     X ← pq.extractMin()
16:     foreach (Y → α) ∈ dep[X] do
17:         cnt[Y → α] ← cnt[Y → α] − 1
18: return ν
```

---

priority queue of nonterminals to "explore" is maintained in the variable pq. The key of each nonterminal in the priority queue is its tentative minimal weight. The lines 2–5 initialize both ν and pq as expected (the tentative minimal weight of each variable is $+\infty$ initially). After this initialization, the algorithm executes the main loop at lines 7–17 until the priority queue is empty. At the beginning of each iteration of the main loop, the priority queue contains the nonterminals whose minimal weight is not determined yet. Formally, it holds at line 7 that $\nu[X] = \mu(X)$ for every $X \in D$, where D is the set $D = V \setminus pq.\texttt{items}()$. Moreover, the variable X contains the last nonterminal that was extracted from the priority queue.[2] First, the **foreach**-loop at lines 9–14 enumerates all productions $Y \to \alpha$ in P such that X occurs(see footnote 2) in α and $\alpha \in (D \cup \Sigma)^*$, and processes them as follows. The weight m arising from the production is computed at line 11. Here, the notation $(\nu \uplus \lambda)$ stands for the function from $V \cup \Sigma$ to $\mathbb{Q}_{\geq 0} \cup \{+\infty\}$ defined by $(\nu \uplus \lambda)(X) = \nu[X]$ and $(\nu \uplus \lambda)(a) = \lambda(a)$, for all $X \in V$ and $a \in \Sigma$. If this weight m is strictly smaller than the tentative minimal weight for Y then both $\nu[Y]$ and the key of Y in the priority queue are updated. Second, at line 15, a nonterminal X with minimum key is extracted from the priority queue. From this point onwards, the tentative minimal weight $\nu[X]$ is

---

[2] Except for the first iteration of the main loop, where $X = \bot$ and where the **foreach**-loop at lines 9–14 processes each $Y \to \alpha$ in P such that α contains no nonterminal.

the actual minimal weight $\mu(X)$. Last, the array *cnt* is updated at lines 16–17 to account for the removal of X from the priority queue.

As with Dijkstra's algorithm, the running time of Algorithm 1 depends on the data structure used to implement the priority queue *pq*. There are $|V|$ `insert` operations, $|V|$ `extractMin` operations, and at most $|P|$ `decreaseKey` operations. Other tasks, including the construction of the arrays *cnt* and *dep* at line 1, run in $O(|V| + t)$ time, where $t = |P| + \sum_{(X \to \alpha) \in P} |\alpha|$ is the *total production length*. Thus, using a binary heap for *pq*, Algorithm 1 runs in $O((|V| + |P|) \log |V| + t)$ time [11]. Fredman and Tarjan improved this time bound to $O(|V| \log |V| + t)$ by using a Fibonacci heap for *pq* [9].

Theorem 1 ([9,11]). *Given a CFG $\mathcal{G} = (V, \Sigma, P)$ and a weight function $\lambda : \Sigma \to \mathbb{Q}_{\geq 0}$, the minimal weight function $\mu : V \to (\mathbb{Q}_{\geq 0} \cup \{+\infty\})$ is computable in $O(|V| \log |V| + t)$ time, where $t = |P| + \sum_{(X \to \alpha) \in P} |\alpha|$.*

4 Shortest Paths in Weighted Interprocedural Automata

This section introduces our generic approach to solve shortest path problems for interprocedural programs through the use of context-free grammars. We start with the definition of our model for (low-level) interprocedural programs.

4.1 Weighted Interprocedural Automata

A *weighted interprocedural automaton*, or *WIPA* for short, is a quintuple $\mathcal{A} = (Q, \Sigma, \mathtt{Op}, \Delta, \lambda)$ where Q is a finite set of *states*, Σ is a finite set of *letters*, $\mathtt{Op} \subseteq \{\tau, \mathtt{ret}\} \cup \{\mathtt{call}\ f \mid f \in Q\}$ is a set of *operations*, $\Delta \subseteq Q \times \Sigma \times \mathtt{Op} \times Q$ is a set of *transitions*, and $\lambda : \Sigma \to \mathbb{Q}_{\geq 0}$ is a *weight function*. A transition $(q, a, \mathtt{op}, q')$ is also written $q \xrightarrow{a,\mathtt{op}} q'$ for readability. It comprises two states $q, q' \in Q$, a letter $a \in \Sigma$, and an operation $\mathtt{op} \in \mathtt{Op}$. The latter is either an *internal operation* (τ), a *function call* ($\mathtt{call}\ f$ with $f \in Q$), or a *function return* ($\mathtt{ret}$). The set of *functions* of $\mathcal{A}$ is $F = \{f \in Q \mid \mathtt{call}\ f \in \mathtt{Op}\}$. An *exit state* is any state $q \in Q$ such that $q \xrightarrow{a,\mathtt{ret}} q' \in \Delta$ for some $a \in \Sigma$ and $q' \in Q$. Exit states correspond to the usual notion of function exit points. Given a function $f \in F$ and a state $q \in Q$, we say that q is a *call state* (resp., *return state*) of f if $q \xrightarrow{a,\mathtt{call}\ f} q' \in \Delta$ (resp., $q' \xrightarrow{a,\mathtt{call}\ f} q \in \Delta$) for some $a \in \Sigma$ and $q' \in Q$. Our running example depicted in Fig. 1 is a WIPA.

The operational semantics of a WIPA $\mathcal{A} = (Q, \Sigma, \mathtt{Op}, \Delta, \lambda)$ is given by the labelled transition system with set of *configurations* $C = Q \times Q^*$ and *step* relation $\hookrightarrow \subseteq C \times \Sigma \times C$ defined by the rules given in Fig. 2. A configuration (q, σ) comprises a state $q \in Q$ and a *stack* content $\sigma \in Q^*$. A *run* of $\mathcal{A}$ is an alternating sequence $\rho = (c_0, a_1, c_1, \ldots, a_n, c_n)$ of configurations $c_i \in C$ and letters $a_i \in \Sigma$ such that $c_{i-1} \xrightarrow{a_i} c_i$ for all $i \in \{1, \ldots, n\}$. We say that ρ is a run *from c_0 to c_n*. The word $a_1 \cdots a_n \in \Sigma^*$ is called the *trace* of ρ. For every word $w \in \Sigma^*$, we let $\xrightarrow{w}$ denote the binary relation on C defined as the set of pairs $(c, c') \in C \times C$

$$\frac{q \xrightarrow{a,\tau} q' \in \Delta \quad \sigma \in Q^*}{(q,\sigma) \overset{a}{\hookrightarrow} (q',\sigma)} \qquad \frac{q \xrightarrow{a,\texttt{call } f} q' \in \Delta \quad \sigma \in Q^*}{(q,\sigma) \overset{a}{\hookrightarrow} (f,q' \cdot \sigma)} \qquad \frac{q \xrightarrow{a,\texttt{ret}} q' \in \Delta \quad r \in Q \quad \sigma \in Q^*}{(q,r \cdot \sigma) \overset{a}{\hookrightarrow} (r,\sigma)}$$

Fig. 2. Operational semantics of a weighted interprocedural automaton.

such that there exists a run from c to c' with trace w. We also write $c \overset{*}{\hookrightarrow} c'$ when $c \overset{w}{\hookrightarrow} c'$ for some $w \in \Sigma^*$. The *distance* from a configuration $c \in C$ to a configuration $c' \in C$ is $\inf\{\lambda(w) \mid c \overset{w}{\hookrightarrow} c'\}$.

Remark 2. The WIPA model, which to our knowledge does not appear in the literature, is similar to the well-known model of *pushdown automata*. The latter uses **push** and **pop** operations while WIPA use **call** and **ret** operations. Our rationale for using WIPA is twofold. First, the direct translation of return statements in low-level (assembly or binary) code requires a **pop** transition to every state of the automaton, whereas a single **ret** transition is needed. Second, WIPA are closer to context-free grammars than pushdown automata, which makes it easier to leverage Theorem 1.

Remark 3. Apart from pushdown automata (see Remark 2), classical models for interprocedural programs are usually given by a collection of pairwise-disjoint automata or graphs, one for each function. On the contrary, in a WIPA, (a) functions are merely specified by their entry state, and (b) several functions may share the same states and transitions. This makes the WIPA model well-adapted to the analysis of low-level (assembly or binary) code, where (a) functions are merely code addresses, and (b) the "body" of a function may contain *goto*-like instructions to the "body" of other functions. This flexibility of the WIPA model slightly complicates its analysis, though.

4.2 Solving Shortest Path Problems Through Context-Free Grammars

The *shortest path problem*, or *SPP* for short, is one of the most fundamental problems in graph theory. Its natural extension to WIPA is the computation of the distances between pairs of states. Distances in a WIPA are only defined between configurations, though. So we consider a number of *stack constraints* on the starting and ending stack contents. Given a stack constraint, the distance from a state q to a state q' is the least distance from (q,σ) to (q',σ') such that (σ,σ') satisfies the considered stack constraint. Each stack constraint induces a variant of the SPP on WIPA. Furthermore, for each of these variants, we investigate both the sources-to-all and the all-to-targets versions of the problem. The *sources-to-all* version is a generalization of the classical single-source version to an arbitrary set $S \subseteq Q$ of *source* states. In the usual setting of finite weighted directed graphs, the sources-to-all version is easily reduced to the single-source version, but this is not the case for some variants of the SPP on WIPA. We shortly refer to the sources-to-all version with set S of sources as $S \triangleright *$. Analogously, the *all-to-targets* version, written $* \triangleright T$, is a generalization of the classical single-target version to an arbitrary set $T \subseteq Q$ of *target* states.

As mentioned in Sect. 1, the SPP on WIPA has many applications in program analysis. In the rest of the paper, we show how to solve all considered variations of the SPP on WIPA, by reduction to the computation of minimal weights in context-free grammars. Our approach is the same in each case, and it can be summarized as follows. Given a WIPA $\mathcal{A} = (Q, \Sigma, \mathsf{Op}, \Delta, \lambda)$, the considered variation of the SPP induces a family $(L_q)_{q \in Q}$ of languages $L_q \subseteq \Sigma^*$. The problem is then the computation of the function from Q to $\mathbb{Q}_{\geq 0} \cup \{+\infty\}$ that maps each state q to $\inf \lambda(L_q)$. To compute this function, we design a CFG $\mathcal{G}$ that contains, for each state $q \in Q$, a nonterminal S_q verifying $L_{S_q}^{\mathcal{G}} = L_q$. The desired function is then computed with Knuth's generalization of Dijkstra's algorithm, see Sect. 3.2.

Note 4. The stated equalities $L_{S_q}^{\mathcal{G}} = L_q$ will hold by construction for each CFG $\mathcal{G}$ that we design in the rest of the paper. Formal proofs of these equalities are mostly routine and rather tedious, so we only provide them in the first two cases (empty stacks and arbitrary stacks). We will present our CFGs in tables such as Table 3. These tables are read line by line as follows. Every instance of the cell in the left-most column induces the productions in the other cells of the line.

Remark 5. The above-mentioned equality $L_{S_q}^{\mathcal{G}} = L_q$ is not strictly necessary for our approach to work. It would be enough to design a CFG $\mathcal{G}$ such that $L_{S_q}^{\mathcal{G}}$ and L_q have the same upward-closure with respect to the subword (subsequence) ordering. This relaxed condition could be exploited to obtain simpler CFGs than the ones presented in this paper.

Empty Stacks. We first consider the case where the stack constraint dictates that both the starting and ending stack contents are empty. The corresponding sources-to-all SPP is, given a WIPA $\mathcal{A} = (Q, \Sigma, \mathsf{Op}, \Delta, \lambda)$ and a set $S \subseteq Q$, the computation of the distance function $D_{S\triangleright *}^{\varepsilon} : Q \to (\mathbb{Q}_{\geq 0} \cup \{+\infty\})$ defined by $D_{S\triangleright *}^{\varepsilon}(q) = \inf \lambda(L_{S\triangleright *}^{\varepsilon}(q))$, where the language $L_{S\triangleright *}^{\varepsilon}(q)$ is given below. In words, $D_{S\triangleright *}^{\varepsilon}(q)$ is the least distance from (s, ε) to (q, ε) where s ranges over the set S of source states. The all-to-targets version of this problem is obtained from the sources-to-all version by replacing S with T and $S \triangleright *$ by $* \triangleright T$.

$$L_{S\triangleright *}^{\varepsilon}(q) = \{w \in \Sigma^* \mid \exists s \in S : (s, \varepsilon) \xrightarrow{w} (q, \varepsilon)\}$$

$$L_{*\triangleright T}^{\varepsilon}(q) = \{w \in \Sigma^* \mid \exists t \in T : (q, \varepsilon) \xrightarrow{w} (t, \varepsilon)\}$$

We solve both versions of the SPP with empty stacks by reduction to the computation of minimal weights in context-free grammars. The corresponding CFGs $\mathcal{G}_{S\triangleright *}^{\varepsilon}$ and $\mathcal{G}_{*\triangleright T}^{\varepsilon}$ are given in Table 2. The set of nonterminals of both CFGs is $\{M_q, S_q \mid q \in Q\}$, and the productions are as defined in the table. To establish the correctness of $\mathcal{G}_{S\triangleright *}^{\varepsilon}$ and $\mathcal{G}_{*\triangleright T}^{\varepsilon}$, we begin with the characterization of the languages of the nonterminals M_q. These nonterminals correspond to matched function calls, and they are used in all CFGs proposed in the paper. Let $\mathcal{M}$ denote the CFG with set of nonterminals $\{M_q \mid q \in Q\}$ and whose productions are as defined in the column *Matched* of Table 2.

Table 2. The context-free grammars $\mathcal{G}^{\varepsilon}_{S\triangleright*}$ and $\mathcal{G}^{\varepsilon}_{*\triangleright T}$ to solve the SPP with empty starting and ending stacks. The grammar $\mathcal{G}^{\varepsilon}_{S\triangleright*}$ uses the columns $S \triangleright *$ and *Matched*, while the grammar $\mathcal{G}^{\varepsilon}_{*\triangleright T}$ uses the columns $* \triangleright T$ and *Matched*.

	$S \triangleright *$	Matched	$* \triangleright T$
$q \xrightarrow{a,\tau} q' \in \Delta$	$S_{q'} \to S_q \cdot a$	$M_q \to a \cdot M_{q'}$	$S_q \to a \cdot S_{q'}$
$q \xrightarrow{a,\texttt{call } f} q' \in \Delta$	$S_{q'} \to S_q \cdot a \cdot M_f$	$M_q \to a \cdot M_f \cdot M_{q'}$	$S_q \to a \cdot M_f \cdot S_{q'}$
$q \xrightarrow{a,\texttt{ret}} q' \in \Delta$		$M_q \to a$	
$q \in S$	$S_q \to \varepsilon$		
$q \in T$			$S_q \to \varepsilon$

Lemma 6. *For every $q \in Q$, $L^{\mathcal{M}}_{M_q}$ is the set of words wa with $w \in \Sigma^*$ and $a \in \Sigma$ such that there exist $q', q'' \in Q$ satisfying $(q, \varepsilon) \xrightarrow{w} (q', \varepsilon)$ and $q' \xrightarrow{a,\texttt{ret}} q'' \in \Delta$.*

The proof of Lemma 6 is routine (see Appendix A for details). We now establish the correctness of $\mathcal{G}^{\varepsilon}_{S\triangleright*}$ and $\mathcal{G}^{\varepsilon}_{*\triangleright T}$. An immediate consequence is that both versions of the SPP with empty stacks can be solved efficiently (see Corollary 8).

Lemma 7. *The CFGs $\mathcal{G}^{\varepsilon}_{S\triangleright*}$ and $\mathcal{G}^{\varepsilon}_{*\triangleright T}$ are computable in $O(|Q| + |\Delta|)$ time. Moreover, for every $q \in Q$, it holds that $L^{\mathcal{G}^{\varepsilon}_{S\triangleright*}}_{S_q} = L^{\varepsilon}_{S\triangleright*}(q)$ and $L^{\mathcal{G}^{\varepsilon}_{*\triangleright T}}_{S_q} = L^{\varepsilon}_{*\triangleright T}(q)$.*

Proof. The time complexity statement follows from the definition of $\mathcal{G}^{\varepsilon}_{S\triangleright*}$ and $\mathcal{G}^{\varepsilon}_{*\triangleright T}$ (see Table 2). We claim that, for every $w \in \Sigma^*$ and $q, q' \in Q$, it holds that $(q, \varepsilon) \xrightarrow{w} (q', \varepsilon)$ iff $S_q \Rightarrow w \cdot S_{q'}$ in $\mathcal{G}^{\varepsilon}_{*\triangleright T}$. Both directions of this claim are routinely shown by induction on $|w|$. The proof is along the same lines as the proof of Lemma 6. The stated equality $L^{\mathcal{G}^{\varepsilon}_{*\triangleright T}}_{S_q} = L^{\varepsilon}_{*\triangleright T}(q)$ follows from this claim and the observation that for every $w \in L^{\mathcal{G}^{\varepsilon}_{*\triangleright T}}_{S_q}$, there exists $q' \in T$ such that $S_q \Rightarrow w \cdot S_{q'}$ in $\mathcal{G}^{\varepsilon}_{*\triangleright T}$. Analogously, the stated equality $L^{\mathcal{G}^{\varepsilon}_{S\triangleright*}}_{S_{q'}} = L^{\varepsilon}_{S\triangleright*}(q')$ derives from, firstly, the claim that $(q, \varepsilon) \xrightarrow{w} (q', \varepsilon)$ iff $S_{q'} \Rightarrow S_q \cdot w$ in $\mathcal{G}^{\varepsilon}_{S\triangleright*}$, and secondly, the observation that for every $w \in L^{\mathcal{G}^{\varepsilon}_{S\triangleright*}}_{S_{q'}}$, there exists $q \in S$ such that $S_{q'} \Rightarrow S_q \cdot w$ in $\mathcal{G}^{\varepsilon}_{S\triangleright*}$. $\quad\square$

Corollary 8. *Both $D^{\varepsilon}_{S\triangleright*}$ and $D^{\varepsilon}_{*\triangleright T}$ are computable in $O(|Q| \log |Q| + |\Delta|)$ time.*

Proof. Observe that $\mathcal{G}^{\varepsilon}_{S\triangleright*}$ and $\mathcal{G}^{\varepsilon}_{*\triangleright T}$ each have $2|Q|$ nonterminals and at most $2|\Delta| + |Q|$ productions (so the total production length is at most $8(|\Delta| + |Q|)$). The corollary directly follows from Lemma 7 and Theorem 1. $\quad\square$

Remark 9. Function summaries can also be computed in $O(|Q| \log |Q| + |\Delta|)$ time, using $\mathcal{G}^{\varepsilon}_{S\triangleright*}$ or $\mathcal{G}^{\varepsilon}_{*\triangleright T}$. In fact, only the productions of the *Matched* column of Table 2 are needed for function summaries (see Lemma 6). The summary of a function $f \in F$ is inf $\lambda(L_f)$, where $L_f = L^{\mathcal{G}^{\varepsilon}_{S\triangleright*}}_{M_f} = L^{\mathcal{G}^{\varepsilon}_{*\triangleright T}}_{M_f}$.

Table 3. The context-free grammar $\mathcal{G}^{\mathrm{arb}}_{S\triangleright*}$ to solve the sources-to-all SPP with arbitrary starting and ending stacks.

	Start	Unmatched returns	Matched	Unmatched calls
$q \xrightarrow{a,\tau} q' \in \Delta$		$R_{q'} \to R_q \cdot a$	$M_q \to a \cdot M_{q'}$	$C_{q'} \to C_q \cdot a$
$q \xrightarrow{a,\mathrm{call}\ f} q' \in \Delta$		$R_{q'} \to R_q \cdot a \cdot M_f$	$M_q \to a \cdot M_f \cdot M_{q'}$	$C_{q'} \to C_q \cdot a \cdot M_f$ $C_f \to C_q \cdot a$
$q \xrightarrow{a,\mathrm{ret}} q' \in \Delta$		$Z \to R_q \cdot a$	$M_q \to a$	
$q \in S$		$R_q \to \varepsilon$		
$q \in Q$	$S_q \to C_q$	$R_q \to Z$		$C_q \to R_q$

Table 4. The context-free grammar $\mathcal{G}^{\mathrm{arb}}_{*\triangleright T}$ to solve the all-to-targets SPP with arbitrary starting and ending stacks.

	Start	Unmatched returns	Matched	Unmatched calls
$q \xrightarrow{a,\tau} q' \in \Delta$		$R_q \to a \cdot R_{q'}$	$M_q \to a \cdot M_{q'}$	$C_q \to a \cdot C_{q'}$
$q \xrightarrow{a,\mathrm{call}\ f} q' \in \Delta$		$R_q \to a \cdot M_f \cdot R_{q'}$	$M_q \to a \cdot M_f \cdot M_{q'}$	$C_q \to a \cdot M_f \cdot C_{q'}$ $C_q \to a \cdot C_f$
$q \xrightarrow{a,\mathrm{ret}} q' \in \Delta$		$R_q \to a \cdot Z$	$M_q \to a$	
$q \in T$				$C_q \to \varepsilon$
$q \in Q$	$S_q \to R_q$	$R_q \to C_q$ $Z \to R_q$		

Arbitrary Stacks. We now consider the case where the stack constraint allows arbitrary starting and ending stack contents. The corresponding sources-to-all and all-to-targets SPPs are the computation of the distance functions $D^{\mathrm{arb}}_{S\triangleright*}(q) = \inf \lambda(L^{\mathrm{arb}}_{S\triangleright*}(q))$ and $D^{\mathrm{arb}}_{*\triangleright T}(q) = \inf \lambda(L^{\mathrm{arb}}_{*\triangleright T}(q))$, respectively, where the languages $L^{\mathrm{arb}}_{S\triangleright*}(q)$ and $L^{\mathrm{arb}}_{*\triangleright T}(q)$ are given below.

$$L^{\mathrm{arb}}_{S\triangleright*}(q) = \{w \in \Sigma^* \mid \exists s \in S, \exists \sigma, \sigma' \in Q^* : (s,\sigma) \xhookrightarrow{w} (q,\sigma')\}$$

$$L^{\mathrm{arb}}_{*\triangleright T}(q) = \{w \in \Sigma^* \mid \exists t \in T, \exists \sigma, \sigma' \in Q^* : (q,\sigma) \xhookrightarrow{w} (t,\sigma')\}$$

Again, we solve both versions of the SPP with arbitrary stacks by reduction to the computation of minimal weights in context-free grammars. The corresponding CFGs $\mathcal{G}^{\mathrm{arb}}_{S\triangleright*}$ and $\mathcal{G}^{\mathrm{arb}}_{*\triangleright T}$ are given in Tables 3 and 4. The set of nonterminals of both CFGs is $\{M_q, S_q, C_q, R_q \mid q \in Q\} \cup \{Z\}$. The nonterminal Z is used to non-deterministically guess the top of the stack. We obtain the following lemma. Its associated corollary again follows from Theorem 1.

Lemma 10. *The CFGs $\mathcal{G}^{\mathrm{arb}}_{S\triangleright*}$ and $\mathcal{G}^{\mathrm{arb}}_{*\triangleright T}$ are computable in $O(|Q| + |\Delta|)$ time. Moreover, for every $q \in Q$, it holds that $L^{\mathcal{G}^{\mathrm{arb}}_{S\triangleright*}}_{S_q} = L^{\mathrm{arb}}_{S\triangleright*}(q)$ and $L^{\mathcal{G}^{\mathrm{arb}}_{*\triangleright T}}_{S_q} = L^{\mathrm{arb}}_{*\triangleright T}(q)$.*

Proof. It is readily seen that for every $w \in \Sigma^*$ and $q, q' \in Q$, if $(q,\sigma) \xhookrightarrow{w} (q',\sigma')$ for some $\sigma, \sigma' \in Q^*$ then $(q,\pi) \xhookrightarrow{x} (p,\varepsilon) \xhookrightarrow{y} (q',\pi')$ for some $p \in Q$, $\pi, \pi' \in Q^*$ and $x, y \in \Sigma^*$ with $w = xy$. We claim that, for every $x, y \in \Sigma^*$ and $p, q, q' \in Q$, the following equivalences hold:

1. $\exists \pi \in Q^* : (q, \pi) \overset{x}{\hookrightarrow} (p, \varepsilon)$ iff $R_p \overset{*}{\Rightarrow} R_q \cdot x$ in $\mathcal{G}^{\mathrm{arb}}_{S \triangleright *}$ iff $R_q \overset{*}{\Rightarrow} x \cdot R_p$ in $\mathcal{G}^{\mathrm{arb}}_{* \triangleright T}$,

2. $\exists \pi' \in Q^* : (p, \varepsilon) \overset{y}{\hookrightarrow} (q', \pi')$ iff $C_{q'} \overset{*}{\Rightarrow} C_p \cdot y$ in $\mathcal{G}^{\mathrm{arb}}_{S \triangleright *}$ iff $C_p \overset{*}{\Rightarrow} y \cdot C_{q'}$ in $\mathcal{G}^{\mathrm{arb}}_{* \triangleright T}$.

These equivalences are routinely shown by induction on $|x|$ and $|y|$. The stated equality $L^{\mathcal{G}^{\mathrm{arb}}_{S \triangleright *}}_{S_{q'}} = L^{\mathrm{arb}}_{S \triangleright *}(q')$ follows from the claim and the observation that for every $w \in L^{\mathcal{G}^{\mathrm{arb}}_{S \triangleright *}}_{S_{q'}}$, there exist $x, y \in \Sigma^*$, $p \in Q$ and $q \in S$ such that $w = xy$ and $S_{q'} \Rightarrow C_{q'} \overset{*}{\Rightarrow} C_p \cdot y \Rightarrow R_p \cdot y \overset{*}{\Rightarrow} R_q \cdot xy$ in $\mathcal{G}^{\mathrm{arb}}_{S \triangleright *}$. The stated equality $L^{\mathcal{G}^{\mathrm{arb}}_{* \triangleright T}}_{S_q} = L^{\mathrm{arb}}_{* \triangleright T}(q)$ follows from the claim and a similar observation on $\mathcal{G}^{\mathrm{arb}}_{* \triangleright T}$. $\square$

Corollary 11. *Both $D^{\mathrm{arb}}_{S \triangleright *}$ and $D^{\mathrm{arb}}_{* \triangleright T}$ are computable in $O(|Q| \log |Q| + |\Delta|)$ time.*

5   Better Handling of Unmatched Returns

In the previous section, we have considered two variants of the shortest path problem on WIPA. In the first variant, the starting stack is empty, which forbids any unmatched return. By *unmatched return*, we mean a function return that is not matched by any function call performed in the run, or, equivalently, that "pops" a state from the starting stack of the run. In the second variant, the starting stack is arbitrary, which gives unmatched returns too much freedom, as they can directly lead to some desired state. Both variants are often unsatisfactory in practice. In this section, we address this issue and present increasingly finer techniques to handle unmatched returns.

The context-free grammars presented in this section are based on $\mathcal{G}^{\mathrm{arb}}_{S \triangleright *}$ and $\mathcal{G}^{\mathrm{arb}}_{* \triangleright T}$ (see Tables 3 and 4). Only the productions regarding unmatched returns are modified.

5.1   Return Edges

Interprocedural control-flow graphs usually come with special edges, called *call edges* and *return edges*, to track the flow of control from caller to callee and vice versa. In this subsection, we propose to use return edges to handle unmatched returns. We first define what we mean by return edges in our setting.

Given a WIPA $\mathcal{A} = (Q, \Sigma, \mathsf{Op}, \Delta, \lambda)$, we define the binary relation $\to$ on Q by $q \to q'$ if $q \overset{a, \mathsf{op}}{\longrightarrow} q' \in \Delta$ for some $a \in \Sigma$ and $\mathsf{op} \in \mathsf{Op}$. Let $\overset{*}{\to}$ denote the reflexive and transitive closure of $\to$. We introduce the *return edge relation* $\rightsquigarrow \subseteq Q \times \Sigma \times Q$ defined by $q \overset{a}{\rightsquigarrow} q'$ if there exists $q'' \in Q$ and a function $f \in F$ such that $q \overset{a, \mathtt{ret}}{\longrightarrow} q'' \in \Delta$, q' is a return state of f and $f \overset{*}{\to} q$ in $\mathcal{A}$. Note that $q \overset{a}{\rightsquigarrow} q'$ implies that q is an exit state. We need to extend the step relation $\hookrightarrow$ of $\mathcal{A}$ to account for return edges. Formally, the *extended step relation* $\hookrightarrow_{\mathrm{edg}} \subseteq C \times \Sigma \times C$ is defined by $(q, \sigma) \overset{a}{\hookrightarrow}_{\mathrm{edg}} (q', \sigma')$ if $(q, \sigma) \overset{a}{\hookrightarrow} (q', \sigma')$ or $q \overset{a}{\rightsquigarrow} q'$ and $\sigma = \sigma' = \varepsilon$. An *extended run* is the same as a run except that it uses the extended step relation $\hookrightarrow_{\mathrm{edg}}$ instead of $\hookrightarrow$. The extension $\overset{w}{\hookrightarrow}_{\mathrm{edg}}$ of $\overset{w}{\hookrightarrow}$ is defined as one would expect.

Table 5. Unmatched returns of the context-free grammars $\mathcal{G}^{edg}_{S\triangleright*}$ and $\mathcal{G}^{edg}_{*\triangleright T}$ to solve the SPP with return edges. The grammar $\mathcal{G}^{edg}_{S\triangleright*}$ (resp., $\mathcal{G}^{edg}_{*\triangleright T}$) also uses the productions from the *Start*, *Matched* and *Unmatched calls* columns of the grammar $\mathcal{G}^{arb}_{S\triangleright*}$ (resp., $\mathcal{G}^{arb}_{*\triangleright T}$), see Tables 3 and 4.

	$\mathcal{G}^{edg}_{S\triangleright*}$	$\mathcal{G}^{edg}_{*\triangleright T}$
$q \xrightarrow{a,\tau} q' \in \Delta$	$R_{q'} \to R_q \cdot a$ $Z_q \to Z_{q'}$	$R_q \to a \cdot R_{q'}$ $Z_{q'} \to Z_q$
$q \xrightarrow{a,call\ f} q' \in \Delta$	$R_{q'} \to R_q \cdot a \cdot M_f$ $R_{q'} \to Z_f$ $Z_q \to Z_{q'}$	$R_q \to a \cdot M_f \cdot R_{q'}$ $Z_f \to R_{q'}$ $Z_{q'} \to Z_q$
$q \xrightarrow{a,ret} q' \in \Delta$	$Z_q \to R_q \cdot a$	$R_q \to a \cdot Z_q$
$q \in S$	$R_q \to \varepsilon$	
$q \in Q$		$R_q \to C_q$

The sources-to-all and all-to-targets SPPs with return edges are the computation of the distance functions $D^{edg}_{S\triangleright*}(q) = \inf \lambda(L^{edg}_{S\triangleright*}(q))$ and $D^{edg}_{*\triangleright T}(q) = \inf \lambda(L^{edg}_{*\triangleright T}(q))$, respectively, where the languages $L^{edg}_{S\triangleright*}(q)$ and $L^{edg}_{*\triangleright T}(q)$ are:

$$L^{edg}_{S\triangleright*}(q) = \{w \in \Sigma^* \mid \exists s \in S, \exists \sigma \in Q^* : (s,\varepsilon) \xrightarrow{w}_{edg} (q,\sigma)\}$$

$$L^{edg}_{*\triangleright T}(q) = \{w \in \Sigma^* \mid \exists t \in T, \exists \sigma \in Q^* : (q,\varepsilon) \xrightarrow{w}_{edg} (t,\sigma)\} \,.$$

It should be clear that $L^{edg}_{S\triangleright*}(q) \subseteq L^{arb}_{S\triangleright*}(q)$ and $L^{edg}_{*\triangleright T}(q) \subseteq L^{arb}_{*\triangleright T}(q)$, for every state $q \in Q$. The reason is that unmatched returns (permitted by extended steps) can be mimicked by matched returns provided that we start with an appropriate stack. Hence, it also holds that $D^{arb}_{S\triangleright*}(q) \leq D^{edg}_{S\triangleright*}(q)$ and $D^{arb}_{*\triangleright T}(q) \leq D^{edg}_{*\triangleright T}(q)$.

We solve both versions of the SPP with return edges by reduction to the computation of minimal weights in context-free grammars. The corresponding CFGs $\mathcal{G}^{edg}_{S\triangleright*}$ and $\mathcal{G}^{edg}_{*\triangleright T}$ are given in Table 5. The set of nonterminals of both CFGs is $\{M_q, S_q, C_q, R_q, Z_q \mid q \in Q\}$. Recall that our CFGs $\mathcal{G}^{arb}_{S\triangleright*}$ and $\mathcal{G}^{arb}_{*\triangleright T}$ for the case with arbitrary stacks used a single nonterminal Z to non-deterministically guess the top of the stack (see Tables 3 and 4). Here, instead of this non-deterministic guess, return edges are discovered and traversed using nonterminals Z_q and productions that explore the WIPA in the reverse direction.

Lemma 12. *The CFGs $\mathcal{G}^{edg}_{S\triangleright*}$ and $\mathcal{G}^{edg}_{*\triangleright T}$ are computable in $O(|Q| + |\Delta|)$ time. Moreover, for every $q \in Q$, it holds that $L^{\mathcal{G}^{edg}_{S\triangleright*}}_{S_q} = L^{edg}_{S\triangleright*}(q)$ and $L^{\mathcal{G}^{edg}_{*\triangleright T}}_{S_q} = L^{edg}_{*\triangleright T}(q)$.*

Corollary 13. *Both $D^{edg}_{S\triangleright*}$ and $D^{edg}_{*\triangleright T}$ are computable in $O(|Q| \log |Q| + |\Delta|)$ time.*

5.2   Coherent Stacks

With return edges, the candidate return states for an exit state q are the return states of the functions f that q belongs to (in the sense that $f \xrightarrow{*} q$). Exactly

Table 6. Unmatched returns of the context-free grammars $\mathcal{G}^{\mathrm{coh}}_{S\triangleright*}$ and $\mathcal{G}^{\mathrm{coh}}_{*\triangleright T}$ to solve the SPP with coherent starting stacks. The grammar $\mathcal{G}^{\mathrm{coh}}_{S\triangleright*}$ (resp., $\mathcal{G}^{\mathrm{coh}}_{*\triangleright T}$) also uses the productions from the *Start*, *Matched* and *Unmatched calls* columns of the grammar $\mathcal{G}^{\mathrm{arb}}_{S\triangleright*}$ (resp., $\mathcal{G}^{\mathrm{arb}}_{*\triangleright T}$), see Tables 3 and 4.

	$\mathcal{G}^{\mathrm{coh}}_{S\triangleright*}$	Matchable	$\mathcal{G}^{\mathrm{coh}}_{*\triangleright T}$
$q \xrightarrow{a,\tau} q' \in \Delta$	$Z_q \to Z_{q'}$	$N_q \to N_{q'}$	$Z_{q'} \to Z_q$
$q \xrightarrow{a,\mathtt{call}\ f} q' \in \Delta$	$R_{q'} \to Z_f$	$N_q \to N_f \cdot N_{q'}$	$Z_f \to R_{q'}$
	$Z_q \to N_f \cdot Z_{q'}$		$Z_{q'} \to Z_q \cdot N_f$
$q \xrightarrow{a,\mathtt{ret}} q' \in \Delta$		$N_q \to \varepsilon$	
$q \in S$	$R_q \to \varepsilon$		
$q \in Q$	$Z_q \to R_q \cdot M_q$		$R_q \to C_q$
			$R_q \to M_q \cdot Z_q$

one such function exists in high-level programs, but this does not hold anymore for low-level programs, as already mentioned in Remark 3. In fact, a WIPA may have exit states that are shared by several functions (even possibly all functions). In that case, an unmatched return may lead to any return state of any of these functions, which is likely to be coarser than what is desired.

To circumvent this issue, we propose to refine the technique of return edges by using a constraint on the starting stack. This technique is inspired from the notion of coherence presented in [5]. Given a WIPA $\mathcal{A} = (Q, \Sigma, \mathtt{Op}, \Delta, \lambda)$, a pair $(q, q') \in Q \times Q$ is called *coherent* if $(q'', \varepsilon) \xhookrightarrow{*} (q, q')$ for some state $q'' \in Q$. An equivalent and maybe more intuitive formulation is that (q, q') is coherent if there exists a function $f \in F$ such that q' is a return state of f and $(f, \varepsilon) \xhookrightarrow{*} (q, \varepsilon)$. By extension, a word $q_1 \cdots q_k \in Q^+$ is called *coherent* if (q_{i-1}, q_i) is coherent for every $i \in \{2, \ldots, k\}$. Finally, a configuration $(q, \sigma) \in C$ is *coherent* if the word $q\sigma$ is coherent. Clearly, every configuration with an empty stack is coherent. We let C^{coh} denote the set of all coherent configurations. Observe that for every run $c \xhookrightarrow{*} c'$ with no unmatched function call, if c is coherent then so is c'.

The sources-to-all and all-to-targets SPPs with coherent stacks are the computation of the distance functions $D^{\mathrm{coh}}_{S\triangleright*}(q) = \inf \lambda(L^{\mathrm{coh}}_{S\triangleright*}(q))$ and $D^{\mathrm{coh}}_{*\triangleright T}(q) = \inf \lambda(L^{\mathrm{coh}}_{*\triangleright T}(q))$, respectively, where the languages $L^{\mathrm{coh}}_{S\triangleright*}(q)$ and $L^{\mathrm{coh}}_{*\triangleright T}(q)$ are:

$$L^{\mathrm{coh}}_{S\triangleright*}(q) = \{w \in \Sigma^* \mid \exists s \in S, \exists \sigma, \sigma' \in Q^* : (s, \sigma) \xhookrightarrow{w} (q, \sigma') \wedge (s, \sigma) \in C^{\mathrm{coh}}\}$$

$$L^{\mathrm{coh}}_{*\triangleright T}(q) = \{w \in \Sigma^* \mid \exists t \in T, \exists \sigma, \sigma' \in Q^* : (q, \sigma) \xhookrightarrow{w} (t, \sigma') \wedge (q, \sigma) \in C^{\mathrm{coh}}\}.$$

Observe that each function return from a coherent configuration can be mimicked by a return edge. This is a consequence of the definition of coherence. It follows that $L^{\mathrm{coh}}_{S\triangleright*}(q) \subseteq L^{\mathrm{edg}}_{S\triangleright*}(q)$ and $L^{\mathrm{coh}}_{*\triangleright T}(q) \subseteq L^{\mathrm{edg}}_{*\triangleright T}(q)$, for every state $q \in Q$. Hence, it also holds that $D^{\mathrm{edg}}_{S\triangleright*}(q) \leq D^{\mathrm{coh}}_{S\triangleright*}(q)$ and $D^{\mathrm{edg}}_{*\triangleright T}(q) \leq D^{\mathrm{coh}}_{*\triangleright T}(q)$.

Again, we solve both versions of the SPP with coherent stacks by reduction to the computation of minimal weights in context-free grammars. The corresponding CFGs $\mathcal{G}_{S\triangleright*}^{\mathrm{coh}}$ and $\mathcal{G}_{*\triangleright T}^{\mathrm{coh}}$ are given in Table 6. The set of nonterminals of both CFGs is $\{M_q, S_q, C_q, R_q, Z_q, N_q \mid q \in Q\}$. The nonterminals Z_q are now used to check coherence. More precisely, it can be shown that for every states $q, q' \in Q$, the pair (q, q') is coherent iff $R_{q'} \overset{*}{\Rightarrow} Z_q$ in $\mathcal{G}_{S\triangleright*}^{\mathrm{coh}}$ iff $Z_q \overset{*}{\Rightarrow} R_{q'}$ in $\mathcal{G}_{*\triangleright T}^{\mathrm{coh}}$. The new nonterminals N_q are required because coherence is based on reachability in the semantics (namely, the condition $(f, \varepsilon) \overset{*}{\hookrightarrow} (q, \varepsilon)$), whereas return edges are based on reachability in the control-flow graph (namely, the condition $f \overset{*}{\rightarrow} q$).

Lemma 14. *The CFGs $\mathcal{G}_{S\triangleright*}^{\mathrm{coh}}$ and $\mathcal{G}_{*\triangleright T}^{\mathrm{coh}}$ are computable in $O(|Q| + |\Delta|)$ time. Moreover, for every $q \in Q$, it holds that $L_{S_q}^{\mathcal{G}_{S\triangleright*}^{\mathrm{coh}}} = L_{S\triangleright*}^{\mathrm{coh}}(q)$ and $L_{S_q}^{\mathcal{G}_{*\triangleright T}^{\mathrm{coh}}} = L_{*\triangleright T}^{\mathrm{coh}}(q)$.*

Corollary 15. *Both $D_{S\triangleright*}^{\mathrm{coh}}$ and $D_{*\triangleright T}^{\mathrm{coh}}$ are computable in $O(|Q| \log |Q| + |\Delta|)$ time.*

5.3 Reachable Stacks

Coherent stacks still suffer from potentially invalid unmatched returns, since coherent configurations are not necessarily reachable. For instance, in our running example of Fig. 1, the configuration (N, KI) is coherent, since the pairs (N, K) and (K, I) are coherent. Yet, the configuration (N, KI) is clearly unreachable, i.e., there is no run from (A, ε) to (N, KI). Indeed, in order to call H, we must call F and not G. This implies that I cannot be in the stack.

We solve this issue by directly requiring that the starting stack is reachable (and, hence, so is the ending stack). This is the most precise stack constraint that is considered in the paper. Of course, reachable stacks are defined with respect to an initial configuration. Therefore, we consider in this subsection WIPA $\mathcal{A} = (Q, \Sigma, \mathrm{Op}, \Delta, \lambda)$ that are additionally equipped with an initial state $q_{init} \in Q$. The initial configuration is then (q_{init}, ε). A configuration $(q, \sigma) \in C$ is called *reachable* when $(q_{init}, \varepsilon) \overset{*}{\hookrightarrow} (q, \sigma)$.

The sources-to-all and all-to-targets SPPs with reachable stacks are the computation of the distance functions $D_{S\triangleright*}^{\mathrm{rea}}(q) = \inf \lambda(L_{S\triangleright*}^{\mathrm{rea}}(q))$ and $D_{*\triangleright T}^{\mathrm{rea}}(q) = \inf \lambda(L_{*\triangleright T}^{\mathrm{rea}}(q))$, respectively, where the languages $L_{S\triangleright*}^{\mathrm{rea}}(q)$ and $L_{*\triangleright T}^{\mathrm{rea}}(q)$ are:

$$L_{S\triangleright*}^{\mathrm{rea}}(q) = \{w \in \Sigma^* \mid \exists s \in S, \exists \sigma, \sigma' \in Q^* : (q_{init}, \varepsilon) \overset{*}{\hookrightarrow} (s, \sigma) \overset{w}{\hookrightarrow} (q, \sigma')\}$$

$$L_{*\triangleright T}^{\mathrm{rea}}(q) = \{w \in \Sigma^* \mid \exists t \in T, \exists \sigma, \sigma' \in Q^* : (q_{init}, \varepsilon) \overset{*}{\hookrightarrow} (q, \sigma) \overset{w}{\hookrightarrow} (t, \sigma')\} \,.$$

We observe that if (q, σ) is reachable and $(q, \sigma) \overset{w}{\hookrightarrow} (q', \sigma')$ then $(q, \pi) \overset{w}{\hookrightarrow} (q', \pi')$ for some π, π' such that (q, π) is coherent (and π is a prefix of σ). This entails that $L_{S\triangleright*}^{\mathrm{rea}}(q) \subseteq L_{S\triangleright*}^{\mathrm{coh}}(q)$ and $L_{*\triangleright T}^{\mathrm{rea}}(q) \subseteq L_{*\triangleright T}^{\mathrm{coh}}(q)$, for every state $q \in Q$. Hence, it also holds that $D_{S\triangleright*}^{\mathrm{coh}}(q) \leq D_{S\triangleright*}^{\mathrm{rea}}(q)$ and $D_{*\triangleright T}^{\mathrm{coh}}(q) \leq D_{*\triangleright T}^{\mathrm{rea}}(q)$. The corresponding CFGs $\mathcal{G}_{S\triangleright*}^{\mathrm{rea}}$ and $\mathcal{G}_{*\triangleright T}^{\mathrm{rea}}$ are given in Table 7. The set of nonterminals of both CFGs is $\{M_q, S_q, C_q, R_q, Z_q, N_q, I_q \mid q \in Q\}$. These two CFGs work differently from the previous ones. Intuitively, the nonterminals I_q and Z_q are now used to determine a prefix of the stack by guessing (on-the-fly) a run from the initial configuration (q_{init}, ε) and tracking the unmatched calls along that run. Only those unmatched calls that are later used by unmatched returns are tracked.

Table 7. Unmatched returns of the context-free grammars $\mathcal{G}^{\text{rea}}_{S\triangleright*}$ and $\mathcal{G}^{\text{rea}}_{*\triangleright T}$ to solve the SPP with reachable starting (and ending) stacks. The grammar $\mathcal{G}^{\text{rea}}_{S\triangleright*}$ (resp., $\mathcal{G}^{\text{rea}}_{*\triangleright T}$) also uses the productions from the *Start*, *Matched* and *Unmatched calls* columns of the grammar $\mathcal{G}^{\text{arb}}_{S\triangleright*}$ (resp., $\mathcal{G}^{\text{arb}}_{*\triangleright T}$), see Tables 3 and 4, as well as the productions from the *Matchable* column of Table 6.

	$\mathcal{G}^{\text{rea}}_{S\triangleright*}$	Reachable	$\mathcal{G}^{\text{rea}}_{*\triangleright T}$
$q \xrightarrow{a,\tau} q' \in \Delta$	$Z_q \to Z_{q'}$	$I_{q'} \to I_q$	$Z_{q'} \to Z_q$
$q \xrightarrow{a,\text{call } f} q' \in \Delta$	$Z_q \to Z_f \cdot M_{q'}$	$I_{q'} \to I_q \cdot N_f$	$Z_f \to M_{q'} \cdot Z_q$
	$Z_q \to N_f \cdot Z_{q'}$	$I_f \to I_q$	$Z_f \to I_q \cdot C_{q'}$
	$R_{q'} \to I_q \cdot Z_f$		$Z_{q'} \to Z_q \cdot N_f$
$q \in S$	$Z_q \to M_q$		
	$R_q \to I_q$		
$q \in Q$			$R_q \to I_q \cdot C_q$
			$R_q \to M_q \cdot Z_q$
q_{init}		$I_q \to \varepsilon$	

Lemma 16. *The CFGs $\mathcal{G}^{\text{rea}}_{S\triangleright*}$ and $\mathcal{G}^{\text{rea}}_{*\triangleright T}$ are computable in $O(|Q| + |\Delta|)$ time. Moreover, for every $q \in Q$, it holds that $L^{\mathcal{G}^{\text{rea}}_{S\triangleright*}}_{S_q} = L^{\text{rea}}_{S\triangleright*}(q)$ and $L^{\mathcal{G}^{\text{rea}}_{*\triangleright T}}_{S_q} = L^{\text{rea}}_{*\triangleright T}(q)$.*

Corollary 17. *Both $D^{\text{rea}}_{S\triangleright*}$ and $D^{\text{rea}}_{*\triangleright T}$ are computable in $O(|Q| \log |Q| + |\Delta|)$ time.*

6 Experimental Results

We assess our novel approach using a sample of 1219 programs from the SV-COMP [2] benchmarks.[3] We chose the SV-COMP as it is open-source and provides a wide variety of programs of various length and with only few dynamic jumps. The latter are not handled properly by our implementation (dynamic jumps are replaced by internal operations). The SV-COMP contains many categories and many similar programs inside each category. We picked only one instance per category. This allows us to cover a wide variety of programs, while limiting the size of our sample. Each program is compiled to pure THUMB-2, with target CPU Cortex-M3 and ARMv7-M architectures. The resulting programs contain on average 80K lines of assembly code and the corresponding WIPAs contain on average 1892 states, among which 44 are return states, and 2092 transitions. Complementary metrics can be found in Table 8. The sources and targets in the programs have been selected randomly. In order to evaluate the degree of precision of each context-free grammar, we compute the number of *refined distances*. Given two grammars A and B, with B corresponding to a refinement of unmatched returns compared to A, we say that the distance of a state q is refined by B if $D^A_{S\triangleright*}(q) < D^B_{S\triangleright*}(q)$ for the sources-to-all version, and $D^A_{*\triangleright T}(q) < D^B_{*\triangleright T}(q)$ for the all-to-targets version. The number of refined distances is the number of states for which this property holds. In Fig. 3, we

---

[3] https://gitlab.com/sosy-lab/benchmarking/sv-benchmarks.

Table 8. Table displaying the minimum, median, average and maximum number of states, transitions and return states in the generated WIPAs.

	Min	Median	Average	Max
Number of states	323	726	1892	152319
Number of transitions	364	812	2092	152744
Number of return states	13	28	44	724

Table 9. Table representing the average, median and maximum of the grammar generation duration, of Knuth's algorithm duration and of the ratio of refined distances per number of states with regard to the previous refinement (only for refined instances). The last line shows the percentage of refined instances.

		$S \triangleright *$				$* \triangleright T$			
		AS	RE	CS	RS	AS	RE	CS	RS
Grammar generation	avg	0.09	0.13	0.13	0.15	0.10	0.12	0.14	0.16
generation duration	med	0.01	0.02	0.02	0.02	0.02	0.02	0.02	0.02
(in seconds)	max	9.24	11	10.29	11	9.00	11	11	12
Knuth algorithm	avg	0.03	0.24	0.03	0.04	0.03	0.02	0.03	0.04
duration (in seconds)	med	0.002	0.003	0.004	0.005	0.004	0.003	0.004	0.006
	max	2.95	2.93	2.08	2.93	2.91	2.87	2.69	3.88
Ratio of refined	avg	-	69%	1.2%	3.2%	-	96%	7.2%	19%
distances w.r.t.	med	-	68%	0.3%	0.3%	-	98%	3.5%	0.9%
previous refinement	max	-	90%	34%	39%	-	100%	58%	75%
Refined instances (%)	-	-	66%	5.3%	12%	-	100%	35%	86%

provide the experimental results based on three metrics: (1) the percentage of *refined instances*, that is the percentage of programs for which at least one distance is refined, (2) the average time required to generate each grammar, and (3) the average execution time of Knuth's algorithm. The graph is divided in two sections corresponding to the sources-to-all and all-to-targets versions of the SPP. Moreover, both sections are divided according to the four stack refinements presented in this paper. The green dots on the graph correspond to the percentage of refined instances by using a grammar compared to using the previous one (i.e., on its left). For instance, the second green dot implies that about 5% of all programs are improved by using coherent stacks over return edges. The scale is on the right and is linear. The bars show the combined time required to generate the corresponding grammar (plain and blue), and the execution time of Knuth's algorithm (hatched and red). The scale is on the left and is linear.

The results presented in Fig. 3 demonstrate the efficiency of our approach, as it averages a sub-second execution time. In practice, it rarely exceeds one second for programs containing less than 100K lines of assembly code. Note that the computation time is widely dominated by the generation of the grammar. This stems from the fact that we have not optimized the grammars nor their generation. For example, in most grammars, the nonterminals S_q can be trivially removed. Moreover, Algorithm 1 could be improved in various ways, as already

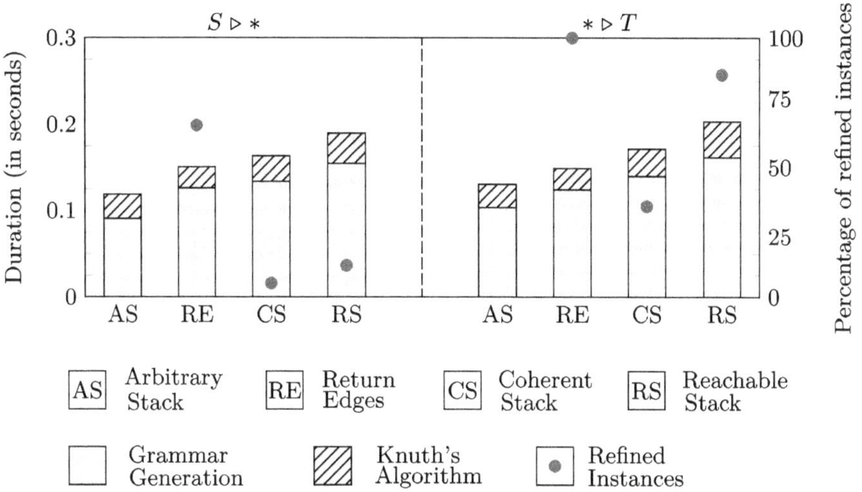

Fig. 3. Experimental results obtained on the SV-COMP benchmarks [2]. Dots represent the percentage of programs for which the distances are more precise, and bars represent the average execution time, in seconds. The plain and hatched sections are, respectively, the duration of the context-free grammar's generation and of Knuth's algorithm. (Color figure online)

noticed in [11]. For instance, we could stop the main `while` loop early, as soon as `extractMin` returns an element with a tentative weight equal to $+\infty$. Indeed, there is no need to compute anything else in that case, as every remaining nonterminal necessarily has a minimal weight of $+\infty$. In Table 9, we provide a textual version of the data shown in Fig. 3. In addition, the medians and maximums are also provided for the same three metrics. Lastly, the table also shows the percentage of refined distances with regard to the number of states. Notice that our refinements of unmatched returns have a greater impact in the all-to-targets version than in the sources-to-all version. We guess that this comes from the asymmetric nature of our refinements (only the starting stack is refined).

On our sample of real low-level programs, using coherent stacks is more precise (especially for the all-to-targets version) and almost as efficient as using return edges. Although slightly less efficient, reachable stacks provide an even greater precision improvement. Besides, the execution time remains low regardless of the refinement used. Experiments also confirm that for every state, return edges are always less precise than, or equivalent to, coherent stacks. The same key observation can be made for coherent stacks against reachable stacks.

7   Conclusion

In this paper, we have tackled several variants (induced by stack constraints) of the shortest path problem for WIPA, and have presented efficient solutions to both the sources-to-all and all-to-targets versions. In each case, we have provided an efficient translation of the problem into a context-free grammar. The solution to the problem is then computed by Knuth's generalization of Dijkstra's algorithm to context-free grammars [11]. The experimental results show that our approach is both efficient and more precise than state-of-the-art methodologies [1,3]. As future work, we intend to optimize our implementation by generating and exploring the grammars on the fly. Another perspective is to address other variants or other versions of the shortest path problem for WIPA.

Acknowledgments. This work was supported by the French ANRT CIFRE 2021/1673 Project.

A   Proof of Lemma 6

Lemma 6. *For every $q \in Q$, $L_{M_q}^{\mathcal{M}}$ is the set of words wa with $w \in \Sigma^*$ and $a \in \Sigma$ such that there exist $q', q'' \in Q$ satisfying $(q, \varepsilon) \overset{w}{\hookrightarrow} (q', \varepsilon)$ and $q' \xrightarrow{a,\mathrm{ret}} q'' \in \Delta$.*

Proof. Let us introduce, for each $q \in Q$, the set L_q of words wa satisfying the lemma condition, formally,

$$L_q = \{ wa \mid w \in \Sigma^*, a \in \Sigma \text{ and } \exists q', q'' \in Q : (q, \varepsilon) \overset{w}{\hookrightarrow} (q', \varepsilon) \wedge q' \xrightarrow{a,\mathrm{ret}} q'' \in \Delta \} .$$

The lemma claims that $L_q = L_{M_q}^{\mathcal{M}}$ for every $q \in Q$.

We show by induction on $|w|$ that $wa \in L_q$ implies $wa \in L_{M_q}^{\mathcal{M}}$, for every $w \in \Sigma^*$, $a \in \Sigma$ and $q \in Q$. For the base case $w = \varepsilon$, we observe that if $a \in L_q$ then there exist $q'' \in Q$ satisfying $q \xrightarrow{a,\mathrm{ret}} q'' \in \Delta$. Hence, $a \in L_{M_q}^{\mathcal{M}}$ since $\mathcal{M}$ contains the production $M_q \to a$. For the induction step, suppose that $(q, \varepsilon) \overset{w}{\hookrightarrow} (q', \varepsilon)$ and $q' \xrightarrow{a,\mathrm{ret}} q'' \in \Delta$ with $|w| > 0$. A run from (q, ε) to (q', ε) either starts with an internal operation or with a function call that is later matched by a function return in the run. So there are exactly two cases to consider.

1. $(q, \varepsilon) \overset{b}{\hookrightarrow} (p, \varepsilon) \overset{x}{\hookrightarrow} (q', \varepsilon)$ for some $b \in \Sigma$, $x \in \Sigma^*$ and $p \in Q$ such that $w = bx$ and $q \xrightarrow{b,\tau} p \in \Delta$. Observe that $xa \in L_p$. By the induction hypothesis, $xa \in L_{M_p}^{\mathcal{M}}$. Note that $\mathcal{M}$ contains the production $M_q \to b \cdot M_p$. It follows that $M_q \Rightarrow b \cdot M_p \overset{*}{\Rightarrow} bxa$ in $\mathcal{M}$, hence, $wa \in L_{M_q}^{\mathcal{M}}$.

2. $(q, \varepsilon) \overset{c}{\hookrightarrow} (f, r) \overset{y}{\hookrightarrow} (p, r) \overset{b}{\hookrightarrow} (r, \varepsilon) \overset{x}{\hookrightarrow} (q', \varepsilon)$ for some $c, b \in \Sigma$, $y, x \in \Sigma^*$ and $f, p, p', r \in Q$ such that $w = cybx$, $q \xrightarrow{c,\mathrm{call}\ f} r \in \Delta$, $(f, \varepsilon) \overset{y}{\hookrightarrow} (p, \varepsilon)$ and $p \xrightarrow{b,\mathrm{ret}} p' \in \Delta$. Observe that $yb \in L_f$ and $xa \in L_r$. By the induction hypothesis, $yb \in L_{M_f}^{\mathcal{M}}$ and $xa \in L_{M_r}^{\mathcal{M}}$. Note that $\mathcal{M}$ contains the production $M_q \to c \cdot M_f \cdot M_r$. It follows that $M_q \Rightarrow c \cdot M_f \cdot M_r \overset{*}{\Rightarrow} cybxa$ in $\mathcal{M}$, hence, $wa \in L_{M_q}^{\mathcal{M}}$.

This concludes the proof of the induction step. We have shown that $L_q \subseteq L_{M_q}^{\mathcal{M}}$ for every $q \in Q$.

Conversely, we start by observing that $|z| \leq |z'|$ for every derivation step $z \Rightarrow z'$ in $\mathcal{M}$. Therefore, $\varepsilon \notin L_{M_q}^{\mathcal{M}}$ for all $q \in Q$. We show by induction on $|w|$ that $wa \in L_{M_q}^{\mathcal{M}}$ implies $wa \in L_q$, for every $w \in \Sigma^*$, $a \in \Sigma$ and $q \in Q$. For the base case $w = \varepsilon$, we derive from the above observation that if $a \in L_{M_q}^{\mathcal{M}}$ then $M_q \Rightarrow a$, hence, $q \xrightarrow{a,\mathtt{ret}} q'' \in \Delta$ for some $q'' \in Q$, and this entails that $a \in L_q$. For the induction step, suppose that $wa \in L_{M_q}^{\mathcal{M}}$ with $|w| > 0$. So we have $M_q \xRightarrow{*} wa$ in $\mathcal{M}$. According to the definition of $\mathcal{M}$, there are exactly two cases to consider for the first derivation step.

1. $M_q \Rightarrow b \cdot M_p \xRightarrow{*} wa$ in $\mathcal{M}$ for some $b \in \Sigma$ and $p \in Q$ such that $q \xrightarrow{b,\tau} p \in \Delta$. As $\varepsilon \notin L_{M_p}^{\mathcal{M}}$, there exists $x \in \Sigma^*$ with $xa \in L_{M_p}^{\mathcal{M}}$ and $wa = bxa$. By the induction hypothesis, $xa \in L_p$, so there exist $q', q'' \in Q$ satisfying $(p,\varepsilon) \xhookrightarrow{x} (q',\varepsilon)$ and $q' \xrightarrow{a,\mathtt{ret}} q'' \in \Delta$. Observe that $(q,\varepsilon) \xhookrightarrow{b} (p,\varepsilon) \xhookrightarrow{x} (q',\varepsilon)$, hence, $(q,\varepsilon) \xhookrightarrow{w} (q',\varepsilon)$. We get that $wa \in L_q$.

2. $M_q \Rightarrow c \cdot M_f \cdot M_r \xRightarrow{*} wa$ in $\mathcal{M}$ for some $c \in \Sigma$ and $f, r \in Q$ such that $q \xrightarrow{c,\mathtt{call}\ f} r \in \Delta$. As $\varepsilon \notin L_{M_f}^{\mathcal{M}}$ and $\varepsilon \notin L_{M_r}^{\mathcal{M}}$, there exist $b \in \Sigma$, $x, y \in \Sigma^*$ with $yb \in L_{M_f}^{\mathcal{M}}$, $xa \in L_{M_r}^{\mathcal{M}}$ and $wa = cybxa$. By the induction hypothesis, $yb \in L_f$ and $xa \in L_r$, so there exist $p, p', q', q'' \in Q$ satisfying $(f,\varepsilon) \xhookrightarrow{y} (p,\varepsilon)$, $p \xrightarrow{b,\mathtt{ret}} p' \in \Delta$, $(r,\varepsilon) \xhookrightarrow{x} (q',\varepsilon)$ and $q' \xrightarrow{a,\mathtt{ret}} q'' \in \Delta$. Observe that $(q,\varepsilon) \xhookrightarrow{c} (f,r) \xhookrightarrow{y} (p,r) \xhookrightarrow{b} (r,\varepsilon) \xhookrightarrow{x} (q',\varepsilon)$, hence, $(q,\varepsilon) \xhookrightarrow{w} (q',\varepsilon)$. We get that $wa \in L_q$.

This concludes the proof of the induction step. We have shown that $L_q \supseteq L_{M_q}^{\mathcal{M}}$ for every $q \in Q$. $\qquad\square$

References

1. Babic, D., Martignoni, L., McCamant, S., Song, D.: Statically-directed dynamic automated test generation. In: Dwyer, M.B., Tip, F. (eds.) Proceedings of the 20th International Symposium on Software Testing and Analysis, ISSTA 2011, Toronto, ON, Canada, 17–21 July 2011, pp. 12–22. ACM (2011)
2. Beyer, D.: State of the art in software verification and witness validation: SV-COMP 2024. In: Finkbeiner, B., Kovács, L. (eds.) Tools and Algorithms for the Construction and Analysis of Systems - 30th International Conference, TACAS 2024, Held as Part of the European Joint Conferences on Theory and Practice of Software, ETAPS 2024, Luxembourg City, Luxembourg, 6–11 April 2024, Proceedings, Part III. Lecture Notes in Computer Science, vol. 14572, pp. 299–329. Springer, Heidelberg (2024). https://doi.org/10.1007/978-3-031-57256-2_15

3. Böhme, M., Pham, V., Nguyen, M., Roychoudhury, A.: Directed greybox fuzzing. In: Thuraisingham, B., Evans, D., Malkin, T., Xu, D. (eds.) Proceedings of the 2017 ACM SIGSAC Conference on Computer and Communications Security, CCS 2017, Dallas, TX, USA, 30 October–03 November 2017, pp. 2329–2344. ACM (2017)
4. Chatterjee, K., Choudhary, B., Pavlogiannis, A.: Optimal Dyck reachability for data-dependence and alias analysis. Proc. ACM Program. Lang. **2**(POPL), 30:1–30:30 (2018)
5. De Castro Pinto, T., Rollet, A., Sutre, G., Tobor, I.: Guiding symbolic execution with A-star. In: Ferreira, C., Willemse, T.A.C. (eds.) Software Engineering and Formal Methods - 21st International Conference, SEFM 2023, Eindhoven, The Netherlands, 6–10 November 2023, Proceedings. Lecture Notes in Computer Science, vol. 14323, pp. 47–65. Springer, Heidelberg (2023). https://doi.org/10.1007/978-3-031-47115-5_4
6. Dijkstra, E.W.: A note on two problems in connexion with graphs. In: Apt, K.R., Hoare, T. (eds.) Edsger Wybe Dijkstra: His Life, Work, and Legacy, ACM Books, vol. 45, pp. 287–290. ACM/Morgan & Claypool (2022)
7. Edelkamp, S., Leue, S., Lluch-Lafuente, A.: Directed explicit-state model checking in the validation of communication protocols. Int. J. Softw. Tools Technol. Transf. **5**(2–3), 247–267 (2004)
8. Fähndrich, M., Rehof, J., Das, M.: Scalable context-sensitive flow analysis using instantiation constraints. In: Lam, M.S. (ed.) Proceedings of the 2000 ACM SIGPLAN Conference on Programming Language Design and Implementation (PLDI), Vancouver, Britith Columbia, Canada, 18–21 June 2000, pp. 253–263. ACM (2000)
9. Fredman, M.L., Tarjan, R.E.: Fibonacci heaps and their uses in improved network optimization algorithms. J. ACM **34**(3), 596–615 (1987)
10. Kjelstrøm, A.H., Pavlogiannis, A.: The decidability and complexity of interleaved bidirected Dyck reachability. Proc. ACM Program. Lang. **6**(POPL), 1–26 (2022)
11. Knuth, D.E.: A generalization of Dijkstra's algorithm. Inf. Process. Lett. **6**(1), 1–5 (1977)
12. Kühnrich, M., Schwoon, S., Srba, J., Kiefer, S.: Interprocedural dataflow analysis over weight domains with infinite descending chains. In: de Alfaro, L. (ed.) Foundations of Software Science and Computational Structures, 12th International Conference, FOSSACS 2009, Held as Part of the Joint European Conferences on Theory and Practice of Software, ETAPS 2009, York, UK, 22–29 March 2009. Proceedings. Lecture Notes in Computer Science, vol. 5504, pp. 440–455. Springer, Heidelberg (2009). https://doi.org/10.1007/978-3-642-00596-1_31
13. Ma, K.-K., Yit Phang, K., Foster, J.S., Hicks, M.: Directed symbolic execution. In: Yahav, E. (ed.) SAS 2011. LNCS, vol. 6887, pp. 95–111. Springer, Heidelberg (2011). https://doi.org/10.1007/978-3-642-23702-7_11
14. Pavlogiannis, A.: CFL/Dyck reachability: an algorithmic perspective. ACM SIGLOG News **9**(4), 5–25 (2022)
15. Rehof, J., Fähndrich, M.: Type-base flow analysis: from polymorphic subtyping to CFL-reachability. In: Hankin, C., Schmidt, D. (eds.) Conference Record of POPL 2001: The 28th ACM SIGPLAN-SIGACT Symposium on Principles of Programming Languages, London, UK, 17–19 January 2001, pp. 54–66. ACM (2001)
16. Reps, T.W., Horwitz, S., Sagiv, S.: Precise interprocedural dataflow analysis via graph reachability. In: Cytron, R.K., Lee, P. (eds.) Conference Record of POPL'95: 22nd ACM SIGPLAN-SIGACT Symposium on Principles of Programming Languages, San Francisco, California, USA, 23–25 January 1995, pp. 49–61. ACM Press (1995)

17. Reps, T.W., Schwoon, S., Jha, S., Melski, D.: Weighted pushdown systems and their application to interprocedural dataflow analysis. Sci. Comput. Program. **58**(1–2), 206–263 (2005)
18. Yang, C.H., Dill, D.L.: Validation with guided search of the state space. In: Chawla, B.R., Bryant, R.E., Rabaey, J.M. (eds.) Proceedings of the 35th Conference on Design Automation, Moscone center, San Francico, California, USA, 15–19 June 1998. pp. 599–604. ACM Press (1998)

0-1 Laws for LTL and CTL over Random Transition Systems

Yanni Dong[1]([✉])(iD), Milan Lopuhaä-Zwakenberg[1](iD), and Mariëlle Stoelinga[1,2](iD)

[1] University of Twente, P.O. Box 217, 7500 AE Enschede, The Netherlands
`yannidong@outlook.com`, `{m.a.lopuhaa,m.i.a.stoelinga}@utwente.nl`
[2] Radboud University, Houtlaan 4, 6525 XZ Nijmegen, The Netherlands

Abstract. This paper examines how 0-1 and convergence laws affect model checking benchmarks, where a formula's truth in a random model becomes nearly independent of the model or follows a specific probability. We investigate these behaviours for Linear Temporal Logic (LTL) and Computation Tree Logic (CTL) in random Kripke structures defined based on Erdős-Rényi random models. For structures with multiple initial states, both LTL and CTL have a 0-1 law. The probability of satisfying an LTL formula converges to 1 or 0 as the model size grows, depending on whether the formula is a tautology or not. In contrast, structures with a single initial state exhibit a convergence law. We also establish that computing asymptotic probabilities for LTL is computationally hard (PSPACE-complete for multiple initial states and PSPACE-hard for one initial state), whereas efficient polynomial-time algorithms exist for CTL. These findings underscore a key limitation: random Kripke structures with multiple initial states are often ineffective as benchmarks because formulae tend to be almost always true or false, irrespective of the model. This renders them unsuitable for evaluating model checking algorithms. Structures with one initial state, however, demonstrate more varied and meaningful behaviour due to convergence laws, where probabilities depend on the model's properties, making them more promising candidates for constructing reliable benchmarks to assess model checking tools.

Keywords: Model checking · Random digraph · Kripke structure · Linear temporal logic · Computation tree logic

1 Introduction

An important aspect of evaluating the performance of model checking techniques is the availability of large-scale benchmarks. However, real-world benchmarks often lack the necessary scale and flexibility to meet these demands. For instance, while benchmarks like the GraBaTs 09 models [39] demonstrate the utility of graph-based approaches to formal verification, they are limited in scalability and configurability [32]. This gap has led to the emergence of random model generation as a promising solution. Techniques such as stochastic parse

G. Ernst and K. Y. Rozier (Eds.): SPIN 2025, LNCS 15945, pp. 66–87, 2026.
https://doi.org/10.1007/978-3-032-06847-7_4

trees for program analysis [19] and probabilistic sampling for Petri net algorithms [16] have been developed to create benchmarks that balance scalability, configurability, and realism while minimizing biases. The Hardware Model Checking Competition (HWMCC) has standardized benchmarks for evaluating industrial problems, using the AIGER format to unify hardware design models. Benchmarks, derived from manual and random generation, assess properties like safety and liveness. Over the past decade, HWMCC has expanded to nearly 2000 models, showcasing field progress. Despite this growth, the need for scalable and flexible benchmarks remains, driving the continued development of random model generation techniques to address these challenges.

A critical question arises: how suitable are randomly generated models for model checking evaluation? Investigating the satisfiability and validity of temporal logic such as Linear Temporal Logic (LTL) or Computation Tree Logic (CTL) formulae on these models is crucial for assessing their utility and advancing the field. To address this question, we must examine the theoretical foundations that govern the behavior of logical formulae in random models, particularly through the lens of asymptotic principles such as 0-1 and convergence laws.

0-1 and Convergence Laws. The behavior of a logic L on large random models can be described by asymptotic principles such as a *0-1 law* or a *convergence law*: A convergence law states that for each $\varphi \in L$, the probability that φ is true in a random model converges to some fixed value c_φ in $[0, 1]$. If furthermore c_φ only takes the values 0 and 1, we say that L adheres to a 0-1 law. Both principles depend on the choice of logic and the method of generating random models.

Relevance of 0-1 and Convergence Laws for Model Checking. 0-1 laws and convergence laws expose fundamental properties of logical systems and their implications for both theoretical and practical applications. While 0-1 laws offer valuable theoretical insights—providing a framework for analyzing the asymptotic behavior and expressive power of logical systems (e.g., if two logics follow a 0-1 law and convergence law over the same graph model, respectively, they must differ in expressive power)—they are often impractical for empirical evaluation. In large models, these laws render most formulae trivially true or false, making it challenging to construct meaningful benchmarks that reflect real-world verification scenarios. Thus, while 0-1 laws help delineate the limitations of logical expressiveness, their direct applicability to practical model checking is limited. In contrast, convergence laws describe scenarios in which the probabilities of formulae stabilize at nontrivial values as the model size increases. These laws are more relevant to practical applications, as they offer a richer and more diverse foundation for evaluating model-checking algorithms. Models governed by convergence laws are better suited for benchmarking.

To investigate the impact of these laws on LTL and CTL model checking in randomly generated Kripke structures, we pose the following question:

Question 1. *Do LTL and CTL follow 0-1 laws, convergence laws, or neither?*

The answer to this question depends on the choice of random model. We consider two types of random Kripke structure, based on the Erdős-Rényi model that is ubiquitous in random graph theory [12,36]. In these models, each transition between states, and each atomic proposition in a state, occurs with a fixed probability. The difference in our two models is in how they handle initial states: we consider a model with multiple initial states, by letting each state be initial with a fixed probability, as well as a model where we uniformly at random choose one from the state set. We specifically focus on one-initial state models, as these are common in certain model checking applications, such as Mealy machines [27], deterministic finite automata [30], and Büchi automata [22]. However, in software and abstract modeling, variables often have multiple possible initial values.

Ignoring some subtleties, random Kripke structures with multiple initial states are an example of what are called random $\mathscr{S}$-structures [14]. These are given by a set V (in our case, the set of states) and some n-ary relations on V; each relation has a fixed probability with which n-ary tuples belong to it. The asymptotic behaviour of $\mathscr{S}$-structures under various logics has been studied extensively [6,9,14,20,21]. In particular, random $\mathscr{S}$-structures follow a 0-1 law under iterative fixed-point logic (IFPL) [3]. Since LTL and CTL are sublogics of IFPL, we directly get the following result:

Answer to Question 1 for Multiple Initial States. For random Kripke structures with multiple initial states, both LTL and CTL follow a 0-1 law.

While this result is known from the literature, the other results of this paper also yield novel, streamlined proofs of this statement that do not require any knowledge of $\mathscr{S}$-structures or IFPL. Regarding the relevance of 0-1 laws for model checking, this demonstrates that this type of random model generation is less effective for evaluating model checkers, as it lacks sensitivity to variations in logical formulae.

One-initial state random Kripke structures are not $\mathscr{S}$-structures, since the initial state cannot be expressed as a relation with fixed probability. As such, existing results cannot be applied, and we require new methods. The following result is a major contribution of this paper:

Answer to Question 1 for One Initial State. For random Kripke structures with one initial state, both LTL and CTL follow a convergence law.

Having settled the existence of a convergence probability c_φ for each LTL/CTL formula φ, the next question is how efficiently c_φ can be obtained (in terms of formula length $|\varphi|$). This question is not considered in previous work regarding $\mathscr{S}$-structures, let alone for one-initial state Kripke structures.

Question 2. *Since the answer to Question 1 is affirmative, can we efficiently compute c_φ given φ?*

Interestingly, the answer for this question differs between LTL and CTL. For LTL, we prove that computing limit probabilities is related to tautology checking, which is known to be PSPACE-complete [34]:

Answer to Question 2 for LTL. For LTL formulae φ, computing c_φ is:

1. PSPACE-complete for multiple-initial state Kripke structures;
2. PSPACE-hard for those with one initial state.

In both cases exponential-time algorithms exist.

For CTL, we find a quadratic-time algorithm to compute c_φ:

Answer to Question 2 for CTL. For CTL formulae, computing the asymptotic probability of satisfaction is polynomial in complexity, regardless of whether the Kripke structures have multiple or one initial state.

These results give the means to determine limit probabilities for large random models, albeit more efficiently for CTL than for LTL. For the single-initial state model, this can be used to obtain formulae with nontrivial limit values; together with large randomly generated models these can be used to flexibly create large benchmarks.

Table 1 summarizes the key findings of this paper. All proofs for these results are provided in the appendix.

Table 1. Asymptotic Behavior of Model Checking for LTL and CTL on Random Kripke Structures: Boldface entries are this paper's main contributions; we also provide new, standalone proofs for the 0-1 laws.

Initial State	Logic	Result	Complexity
Multiple	LTL	0-1 law [3]	**PSPACE-complete**
Multiple	CTL	0-1 law [3]	**Polynomial**
One	LTL	**Convergence Law**	**PSPACE-hard**
One	CTL	**Convergence Law**	**Polynomial**

1.1   Related Work

Erdős–Rényi Random Graphs. Random graph theory is a vibrant field pioneered by Erdős and Rényi [8]. A *directed graph* (or *digraph*) is a graph in which edges have a direction, meaning each edge is an ordered pair of vertices. An Erdős-Rényi random directed graph model $D_{n,p}$ with self-loops has n vertices and each of the n^2 possible edges occurs independently with probability p. Erdős-Rényi random undirected graphs also exist [11]. The study of Erdős-Rényi random (di)graphs has been concerned with whether large random graphs satisfy certain graph properties such as the existence of triangles, vertex connectivity, hamiltonicity and rigidity [4,11,35]. The limit behaviour of large random graphs w.r.t. some graph properties is often stated in terms of 0-1 laws. Convergence laws, where the limit probability exists but is not necessarily 0 or 1, are also studied. We refer the interested readers to the book [36] for such laws.

First-Order Logic and its Extensions. Other researchers explore the link between logical definability and asymptotic probability. Fagin [9] and Glebskii et al. [13] were the first to establish this connection, proving that first-order logic (FOL) follows a 0-1 law [9,13] for constant probability. This finding initiated further research into the relationship between logic expressibility and asymptotic probabilities. However, determining whether the asymptotic probability of an FOL property is 0 or 1 is PSPACE-complete [14]. Additionally, whether FOL obeys the 0-1 law depends on edge probability. Lynch [25] showed that FOL has a convergence law in certain probability cases, while Razafimahatratra and Zhukovskii [31] proved in 2020 that FOL lacks a 0-1 law in some cases. Despite these findings, FOL has limited expressive power. Blass and Harary [4] proved that no FOL sentence with asymptotic probability 1 can imply hamiltonicity or rigidity, though their asymptotic probability is 1. This led to the question of whether a more expressive logic exists. Dawar and Grädel [7] showed that no regular logic (in the model-theoretic sense) can express hamiltonicity. While second-order logic can express hamiltonicity, it does not obey a 0-1 law [23].

In [24], the authors stated that Talanov [37], Talanov and Knyazev [38] showed that transitive closure logic and least fixed point logic have a 0-1 law[1]. Blass, Gurevich, and Kozen [3] proved that a 0-1 law exists for IFPL, which, in particular, includes the corresponding result for the least fixed-point operator. The decision problem of the corresponding logic is EXPTIME-complete [3]. Kolaitis and Vardi [23] announced that partial fixed point logic has a 0-1 law. The decision problem of the corresponding logic is EXPSPACE-complete for a bounded vocabulary and 2EXPSPACE for an unbounded vocabulary [23].

For some recent research on 0-1 and convergence laws, we refer the reader to [15,17,26,33].

2   Random Kripke Structures

In this section, we define random Kripke structures and show how they arise from (random) labelled digraphs.

Definition 1 (Labelled Digraphs and Kripke Structures). *Let k be a positive integer. A* labelled directed graph (digraph) *is a tuple $D = (V, R, I, L_1, \ldots, L_k)$ where V is a finite set, $R \subseteq V \times V$, $I \subseteq V$ and $L_i \subseteq V$ for $i = 1, \ldots, k$. A labelled digraph is called a* Kripke structure *if $I \neq \varnothing$ and R is left-total, i.e., for all $v \in V$ there is a $v' \in V$ such that $(v, v') \in R$.*

Kripke structures model computer programs as follows: the vertices of the directed graph (V, R) represent states of the program, and the edges represent transitions (aka steps) that can be taken in its runtime; thus an execution of a program is a $path^2$, i.e., an infinite sequence of states connected by transitions, in this digraph. Here, I is the set of initial states. Thus, to say that a labelled digraph is a Kripke structure means that initial states exist and that every state

---

[1] We cite [37,38] in this way because we could not reach these papers.

[2] Note that this is different from the usual notion in graph theory, where a path is a finite sequence of edges connecting distinct vertices.

has an outgoing transition (possibly to itself). A Kripke structure is typically equipped with a set of *atomic propositions* $\mathsf{AP} = \{a_1, \ldots, a_k\}$, which represent state properties. Each atomic proposition a_i corresponds to a set of states $L_i \subseteq S$ where it holds, meaning that a state v satisfies a_i (denoted $v \models a_i$) if and only if $v \in L_i$. Since a state can be labelled with multiple atomic propositions, it can belong to multiple L_i sets. The purpose of introducing these sets is to formalize the labelling of states with atomic propositions in a way that facilitates logical reasoning. Logics on Kripke structures typically describe how atomic propositions hold throughout the program's execution, as we will discuss in the next three sections.

For any model checking terminology and notation not covered in this paper, we refer readers to reference [1,18] for further details. For consistency, states and transitions in Kripke structures are called vertices and edges, respectively.

2.1 Random Labelled Digraphs and Kripke Structures

Before we define random Kripke structures we first define random labelled digraphs in general. The key idea is that we assign fixed probabilities to edges existing, vertices being initial, and vertices satisfying atomic propositions; and these events are independent. This construction extends the concept of Erdős-Rényi random graphs [8] to labelled digraphs by considering directed edges and adding labels on vertices.

Definition 2 (Random Labelled Digraphs). *Let $n, k \in \mathbb{N}_{\geq 1}$, and let $\boldsymbol{p} = (p_R, p_I, p_1, \ldots, p_k) \in (0,1)^{k+2}$. Then the* random labelled digraph

$$\mathbb{D}_{n,\boldsymbol{p}}^{+} = (V, R, I, L_1, \ldots, L_k)$$

is defined as:

- *V is a set of vertices with $|V| = n$.*
- *for each $v, v' \in V$ we have $\mathbf{P}((v, v') \in R) = p_R$.*
- *for each $v \in V$ and $i = 1, \ldots, k$ we have $\mathbf{P}(v \in L_i) = p_i$.*
- *for each $v \in V$ we have $\mathbf{P}(v \in I) = p_I$.*

Further, all these events are independent. In other words, we have

$$\mathbf{P}(\mathbb{D}_{n,\boldsymbol{p}}^{+} = (V, R, I, L_1, \ldots, L_k))$$

$$= p_R^{|R|} (1 - p_R)^{n^2 - |R|} p_I^{|I|} (1 - p_I)^{n - |I|} \prod_{i=1}^{k} p_i^{|L_i|} (1 - p_i)^{n - |L_i|}.$$

Note that random labelled digraphs are a special case of $\mathscr{S}$-structures on a universe V, defined in [14], where $\mathscr{S} = \{R, I, L_1, \ldots, L_k\}$ is a set of relations, R is a binary relation, and I and L_is are unary relations.

Our probability distribution on Kripke structures is the same as those on labelled digraphs, conditioned to the fact that they should be Kripke structures.

Definition 3 (Random Kripke Structures with Multiple Initial Vertices). *Let $n, k \in \mathbb{Z}_{\geq 1}$ and let $\boldsymbol{p} = (p_R, p_I, p_1, \ldots, p_k) \in (0,1)^{k+2}$. Then the random Kripke structure with multiple initial vertices $\mathbb{K}_{n,\boldsymbol{p}}^{+}$ is defined by*

$$\mathbf{P}(\mathbb{K}_{n,\boldsymbol{p}}^{+} = K) = \frac{\mathbf{P}(\mathbb{D}_{n,\boldsymbol{p}}^{+} = K)}{\mathbf{P}(\mathbb{D}_{n,\boldsymbol{p}}^{+} \text{ is a Kripke structure})}$$

for all Kripke structures K.

Example 1. Suppose $n = 2$ and $k = 1$, meaning there are two vertices and $AP = \{a_1\}$. Since each of the two vertices can have an edge to itself or the other vertex, there are $2^4 = 16$ possible edge relations R. Additionally, there are $2^2 = 4$ possible choices for both the set of initial states I and the labeling set L_1. Therefore, there are $2^8 = 256$ possible labelled digraphs on two vertices when $k = 1$. Among these, 108 satisfy the conditions of a Kripke structure. Furthermore, it turns out that 57 of these Kripke structures satisfy the CTL formula $\mathsf{AF}a_1$. Now suppose $p_R = p_I = p_1 = \frac{1}{2}$. Under this assumption, $\mathbb{D}_{2,\boldsymbol{p}}^{+}$ follows a uniform distribution over the 256 possible labelled digraphs, and $\mathbb{K}_{2,\boldsymbol{p}}^{+}$ follows a uniform distribution over the 108 Kripke structures. Consequently, the probability that a randomly selected Kripke structure satisfies $\mathsf{AF}a_1$ is given by $\mathbf{P}(\mathbb{K}_{2,\boldsymbol{p}}^{+} \models \mathsf{AF}a_1) = \frac{57}{108}$.

Note that not all 256 possible labelled digraphs in the example above qualify as valid Kripke structures, as they must meet two key conditions: (1) at least one initial state must exist, and (2) every state must have at least one outgoing transition.

As mentioned in the introduction, many model checking applications consider Kripke structures with one initial vertex. In this paper, we show that their behaviour is different from Kripke structures with multiple initial vertices.

Definition 4 (Random Kripke Structure with One Initial Vertex). *Let $n, k \in \mathbb{N}_{\geq 1}$, and let $\hat{\boldsymbol{p}} = (p_R, p_1, \ldots, p_k) \in (0,1)^{k+1}$. The random Kripke structure with one initial vertex, denoted by $\mathbb{K}_{n,\hat{\boldsymbol{p}}}^{1}$, is defined by*

$$\mathbf{P}(\mathbb{K}_{n,\hat{\boldsymbol{p}}}^{1} = K) = \frac{\mathbf{P}(\mathbb{D}_{n,\hat{\boldsymbol{p}}}^{1} = K)}{\mathbf{P}(\mathbb{D}_{n,\hat{\boldsymbol{p}}}^{1} \text{ is a Kripke structure})}$$

for all Kripke structures K with one initial vertex. Here $\mathbb{D}_{n,\hat{\boldsymbol{p}}}^{1} = (V, R, I, L_1, \ldots, L_k)$ is a random labelled digraph with I a singleton, chosen uniformly at random from V.

The following result shows that $\mathbb{D}_{n,\boldsymbol{p}}^{+}$ and $\mathbb{D}_{n,\hat{\boldsymbol{p}}}^{1}$ are almost always Kripke structures; this means that for our purposes, we can often treat $\mathbb{D}_{n,\boldsymbol{p}}^{+}$ and $\mathbb{K}_{n,\boldsymbol{p}}^{+}$ to be identical. The same applies to $\mathbb{D}_{n,\hat{\boldsymbol{p}}}^{1}$ and $\mathbb{K}_{n,\hat{\boldsymbol{p}}}^{1}$.

Proposition 1.

$$\lim_{n \to \infty} \mathbf{P}(\mathbb{D}_{n,\boldsymbol{p}}^{+} \text{ is a Kripke structure}) = \lim_{n \to \infty} \mathbf{P}(\mathbb{D}_{n,\hat{\boldsymbol{p}}}^{1} \text{ is a Kripke structure}) = 1.$$

3   0-1 and Convergence Laws for Propositional Logic

3.1   Results for Propositional Logic

Before diving into more advanced logics, we discuss our results for propositional logic. These are not deep, but set the stage for LTL and CTL.

Consider the set $AP = \{a_1, \ldots, a_k\}$ of atomic propositions. In propositional logic, the syntax of well-formed formulae is given by the following grammar:

$$\alpha ::= \mathit{true} \mid \mathit{false} \mid a_i \mid \neg\alpha \mid \alpha \vee \alpha \mid \alpha \wedge \alpha.$$

For a Kripke structure K with vertex v, we say $v \models a_i$ if $v \in L_i$; we extend this to define $v \models \alpha$ for any propositional formula α. We furthermore say $K \models \alpha$ if $v \models \alpha$ for all initial v. Note that this coincides with both LTL and CTL.

We can consider the a_i to be independent Bernoulli random variables, each with probability p_i. Under this distribution, we define $\pi(\alpha) = \mathbf{P}(\alpha \text{ holds})$.

In this notation we have $\mathbf{P}(v \models \alpha) = \pi(\alpha)$ for each $v \in \mathbb{K}_{n,p}^+$. Hence $\mathbf{P}(\mathbb{K}_{n,\hat{p}}^1 \models \alpha) = \pi(\alpha)$; in particular, propositional logic follows a convergence law on $\mathbb{K}_{n,\hat{p}}^1$. For $\mathbb{K}_{n,p}^+$, it has approximately np_I initial vertices, so we have

$\mathbf{P}(\mathbb{K}_{n,p}^+ \models \alpha) \approx \pi(\alpha)^{np_I}$. Thus $\mathbf{P}(\mathbb{K}_{n,p}^+ \models \alpha) \xrightarrow{\infty} 0$ for non-tautologies[3]; when α is a tautology, we have $\mathbf{P}(\mathbb{K}_{n,p}^+ \models \alpha) \xrightarrow{\infty} 1$. We conclude that propositional logic follows a 0-1 law on $\mathbb{K}_{n,p}^+$.

Computing $\mathbf{P}(\mathbb{K}_{n,\hat{p}}^1 \models \alpha) = \pi(\alpha)$, and determining whether α is a tautology to compute $\lim_{n \to \infty} \mathbf{P}(\mathbb{K}_{n,p}^+ \models \alpha)$, boils down to examining the truth table of α and determining which rows evaluate to 1. Evaluating one row has time complexity $\mathcal{O}(|\alpha|)$; since k is fixed, it is linear in $|\alpha|$. Thus we have the following result:

Proposition 2 (0-1, Convergence, Complexity for Propositional Logic).

1. *Propositional logic follows a 0-1 law on $\mathbb{K}_{n,p}^+$.*
2. *Propositional logic follows a convergence law on $\mathbb{K}_{n,\hat{p}}^1$.*
3. *Determining $\lim_{n\to\infty} \mathbf{P}(\mathbb{K}_{n,p}^+ \models \alpha)$ and $\lim_{n\to\infty} \mathbf{P}(\mathbb{K}_{n,\hat{p}}^1 \models \alpha)$ has time complexity $\mathcal{O}(|\alpha|)$.*

3.2   Extension Statement

The final result of this section is the analog for the well-known extension statement for random graphs [4] to random Kripke structures; it is the key ingredient in many of our results. Informally, the extension statement says that with high probability, any extension of any substructure of $\mathbb{K}_{n,p}^+$ also exists within $\mathbb{K}_{n,p}^+$.

Definition 5. *Let D be a labelled digraph.*

---

[3] For convenience, we write $g(n) \xrightarrow{\infty} r$ for $\lim_{n\to\infty} g(n) = r$.

1. *A labelled sub-digraph of D is a labelled digraph $D' = (V', R', I', L'_1, \ldots, L'_k)$ with $V' \subseteq V$, $R' \subseteq R \cap (V' \times V')$, $I' = I \cap V'$, and $L'_i = L_i \cap V'$ for each i.*
2. *A sub-Kripke structure is a labelled sub-digraph that is a Kripke structure.*
3. *An induced labelled sub-digraph is a labelled sub-digraph where the $\subseteq$ for R' is equality.*
4. *Let D_1 be another labelled digraph. An induced sub-digraph isomorphism $f_1 \colon D_1 \to D$ is an isomorphism from D_1 to an induced labelled sub-digraph of D.*

The following extension statement is proven analogously to the statement on random graphs in [4].

Theorem 1 (Extension Statement). *Let D_1 and D_2 be labelled digraphs where D_1 is an induced labelled sub-digraph of D_2. Then the probability of $\mathbb{K}^+_{n,\boldsymbol{p}}$ having the following property goes to 1 as $n \to \infty$: Every induced sub-digraph isomorphism $f_1 \colon D_1 \to \mathbb{K}^+_{n,\boldsymbol{p}}$ can be extended to an induced sub-digraph isomorphism $f_2 \colon D_2 \to \mathbb{K}^+_{n,\boldsymbol{p}}$.*

Example 2. Take D_1 a vertex v with $v \in I$ and $L_1 = R = \varnothing$. Take D_2 with $V = \{v, v'\}$, $I = \{v\}$, $L_1 = \{v'\}$ and $R = \{(v, v')\}$. Suppose that there exists $f_1 : D_1 \to \mathbb{K}^+_{n,\boldsymbol{p}}$. Actually this assumption is almost surely true. For a vertex in $\mathbb{K}^+_{n,\boldsymbol{p}}$, it is an initial vertex, has no self-loop and is not in L_1 with probability $(1 - p_R)p_I(1 - p_1)$. Furthermore, these events are independent for different vertices, the probability that there exists such a vertex goes to 1 as $n \to \infty$. According to Theorem 1, the asymptotic probability of extending f_1 to $f_2 : D_2 \to \mathbb{K}^+_{n,\boldsymbol{p}}$ is 1, see Fig. 1.

Note that the extension statement is not true for $\mathbb{K}^1_{n,\hat{\boldsymbol{p}}}$: if $D_1 = \varnothing$ and D_2 has two initial vertices, then the unique induced sub-digraph isomorphism $D_1 \to \mathbb{K}^1_{n,\hat{\boldsymbol{p}}}$ can never be extended to D_2.

4   0-1 and Convergence Laws for LTL

LTL is a temporal logic that utilizes atomic propositions along with operators such as negation, conjunction, disjunction, and temporal operators X (Next), F (Finally), U (Until), G (Always) to specify properties of linear paths. In this section, we present our results on LTL.

4.1   0-1 Law for Multiple Initial Vertices

In this section we prove that LTL follows a 0-1 law for $\mathbb{K}^+_{n,\boldsymbol{p}}$. While this has already been proven in [4], our proof is self-contained as it does not rely on intermediate fixed-point logic. Furthermore, our proof also immediately gives the limit probability for each φ. More precisely, the limit probability is 1 iff φ is a tautology. The intuitive reason is that if φ is not a tautology, there exists a counterexample and a path in that counterexample for which φ does not hold. Using the extension statement, we can show that this path exists in $\mathbb{K}^+_{n,\boldsymbol{p}}$ with high probability, hence $\mathbb{K}^+_{n,\boldsymbol{p}} \not\models \varphi$ with high probability. The next example illustrates this.

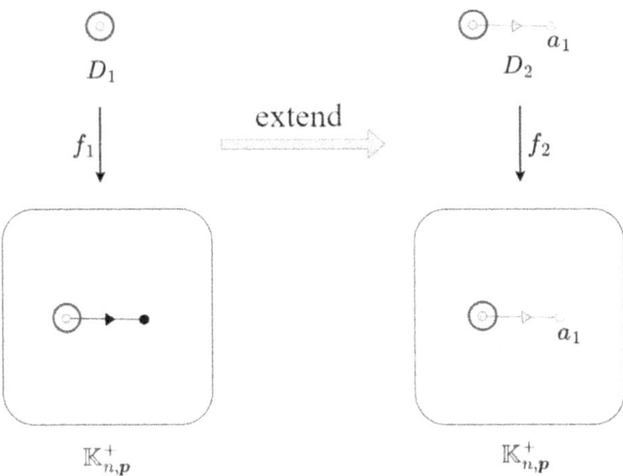

Fig. 1. Explanation Examples 2 and 5: This explanation describes the extension of Example 2 and one possible extension of Example 5. The vertices inside the large thick circle represent the initial vertices. We use a shadowed box to abstract it into the remaining part of $\mathbb{K}_{n,p}^{+}$. Vertices and edges sharing the same vibrant color indicate their correspondence before and after the maps.

Example 3. Consider LTL formulae $\mathsf{F}a_1$ and $\mathsf{G}a_1$. With asymptotic probability 1, there exists a path π^* in $\mathbb{K}_{n,p}^{+}$ such that $\pi^* \not\models \mathsf{F}a_1$ and $\pi^* \not\models \mathsf{G}a_1$. For example, a path of the form $v, v, v, v, \ldots$ satisfies this condition when v is a vertex in $\mathbb{K}_{n,p}^{+}$ satisfying $v \in I, v \notin L_1$ and $(v, v) \in R$. By Theorem 2, we prove almost surely that there exists a vertex v in $\mathbb{K}_{n,p}^{+}$ such that $v \in I, v \notin L_1$ and $(v, v) \in R$ by applying $D_1 = \varnothing$ and D_2 is such a vertex. Thus $\lim_{n\to\infty} \mathbf{P}(\mathbb{K}_{n,p}^{+} \models \mathsf{F}a_1) = \lim_{n\to\infty} \mathbf{P}(\mathbb{K}_{n,p}^{+} \models \mathsf{G}a_1) = 0$.

This insight leads us to the key ingredient in the following lemma, which follows directly from Theorem 1 by taking $D_1 = \varnothing$ and $D_2 = K$.

Lemma 1. *Let K be a Kripke structure. Then*

$$\lim_{n\to\infty} \mathbf{P}(\mathbb{K}_{n,p} \text{ contains } K \text{ as an induced sub-Kripke structure}) = 1.$$

This allows us to prove the following result, amounting to a 0-1 law for LTL:

Theorem 2 (0-1 Law for LTL). *Let φ be an LTL formula. Then we have $\mathbf{P}(\mathbb{K}_{n,p}^{+} \models \varphi) \xrightarrow{\infty} 1$ if φ is a tautology, and $\mathbf{P}(\mathbb{K}_{n,p}^{+} \models \varphi) \xrightarrow{\infty} 0$ otherwise.*

The idea of the proof is as follows: clearly if φ is a tautology, then $\mathbf{P}(\mathbb{K}_{n,p}^{+} \models \varphi) \xrightarrow{\infty} 1$. If φ is not a tautology, then there is a Kripke structure K such that $K \not\models \varphi$. There exists a path π in K such that $K, \pi \not\models \varphi$. By Lemma 1, $\mathbb{K}_{n,p}^{+}$ contains K almost surely. This implies that $\mathbb{K}_{n,p}^{+}$ contains the path π almost surely, so almost surely $\mathbb{K}_{n,p}^{+}, \pi \not\models \varphi$, that is $\mathbb{K}_{n,p}^{+} \not\models \varphi$.

Note that Theorem 2 is independent of the choice of $\boldsymbol{p}$. Moreover, $\mathbb{K}^1_{n,\hat{\boldsymbol{p}}}$ does not satisfy a 0-1 law since $\mathbf{P}(\mathbb{K}^1_{n,\hat{\boldsymbol{p}}} \models a_1) = p_1$ does not converge to 0 or 1. However, in the next subsection, we demonstrate that LTL follows a convergence law for $\mathbb{K}^1_{n,\hat{\boldsymbol{p}}}$.

4.2   Case 2: One Initial Vertex

To formulate the limit behaviour of LTL on $\mathbb{K}^1_{n,\hat{\boldsymbol{p}}}$ we need some extra notation. Recall that $AP = \{a_1, \ldots, a_k\}$ is the set of atomic propositions. A *minterm* is a conjunction of literals in which each variable is represented exactly once. Let $\mathcal{S}$ be the set of minterms:

$$\mathcal{S} = \left\{ \bigwedge_{i=1}^{k} \ell_i \;\middle|\; \ell_i \in \{a_i, \neg a_i\} \right\}.$$

Clearly, for each vertex v in a Kripke structure there is precisely one $\sigma \in \mathcal{S}$ such that $v \models \sigma$. Let $\mathbb{K}^1_{n,\hat{\boldsymbol{p}}}(\sigma)$ denote one random Kripke structure $\mathbb{K}^1_{n,\hat{\boldsymbol{p}}}$ conditioned on the properties that $\mathbb{K}^1_{n,\hat{\boldsymbol{p}}} \models \sigma$ (i.e., the single initial vertex satisfies σ). An *initial self-loop* is an edge that connects the initial vertex to itself.

We then get the following analogue of Lemma 1:

Lemma 2. *Let $\sigma \in \mathcal{S}$, and let K be a Kripke structure with one initial vertex and without initial self-loop such that $K \models \sigma$. Then*

$$\lim_{n \to \infty} \mathbf{P}(\mathbb{K}^1_{n,\hat{\boldsymbol{p}}}(\sigma) \ contains \ K \ as \ a \ sub\text{-}Kripke \ structure) = 1.$$

We cannot guarantee whether $\mathbb{K}^1_{n,\hat{\boldsymbol{p}}}(\sigma)$ includes an initial self-loop. If it lacks an initial self-loop, then it cannot contain any Kripke structure with an initial self-loop as a sub-Kripke structure. This is one reason why, in Lemma 2, we consider a Kripke structure without initial self-loop. However, as we will see later, this distinction does not affect the overall analysis. The proof of Lemma 2 again relies on Theorem 1, though we need some subtlety to deal with the self-loop, which $\mathbb{K}^1_{n,\hat{\boldsymbol{p}}}(\sigma)$ may or may not have. The sub-Kripke structure involved in the above result is influenced by the propositional formula satisfied by the initial vertex of the random Kripke structure. In contrast, Lemma 1 is not affected in this way because, as n approaches infinity, there are sufficiently many initial vertices in $\mathbb{K}^+_{n,\boldsymbol{p}}$ satisfying various propositional formulae.

Lemma 3. *Let φ be an LTL formula. Then we have $\mathbf{P}(\mathbb{K}^1_{n,\hat{\boldsymbol{p}}}(\sigma) \models \varphi) \xrightarrow{\infty} 1$ if $\sigma \to \varphi$ is a tautology, and $\mathbf{P}(\mathbb{K}^1_{n,\hat{\boldsymbol{p}}}(\sigma) \models \varphi) \xrightarrow{\infty} 0$ otherwise.*

The idea of the proof is as follows. If $\sigma \to \varphi$ is a tautology, then clearly $\lim_{n \to \infty} \mathbf{P}(\mathbb{K}^1_{n,\hat{\boldsymbol{p}}}(\sigma) \models \varphi) = 1$. Suppose that $\sigma \to \varphi$ is not a tautology. We prove $\mathbf{P}(\mathbb{K}^1_{n,\hat{\boldsymbol{p}}}(\sigma) \models \varphi) \xrightarrow{\infty} 0$. First, we claim that there is a counterexample of $\sigma \to \varphi$ with one initial vertex and without initial self-loop. Since $\sigma \to \varphi$ is not a tautology, there exists a Kripke structure K with a path π such that

$K, \pi \not\models \sigma \to \varphi$; we may assume the initial vertex of π is the only initial vertex. The fact that $K, \pi \not\models \sigma \to \varphi$ now means that $K \models \sigma$ and $K, \pi \not\models \varphi$.

To apply Lemma 2, we need a counterexample without initial self-loop. If K has a self-loop, we can create another counterexample K' by taking two copies of K and replacing these self-loops with edges from one copy to the other (see Fig. 2). Then π has a unique lift to a path π' in K', and $K', \pi' \not\models \sigma \to \varphi$. Furthermore, K' has no initial self-loop. Either way, Lemma 2 shows us that the counterexample for φ also exists in $\mathbb{K}^1_{n,\hat{p}}$ with high probability. As in Theorem 2, this implies $\mathbf{P}(\mathbb{K}^1_{n,\hat{p}}(\sigma) \models \varphi) \xrightarrow{\infty} 0$.

Fig. 2. Constructing a Kripke Structure Without the Initial Self-Loop: This example demonstrates the transformation of a Kripke structure that initially has one initial vertex with the initial self-loop into a structure with one initial vertex and no initial self-loop. The vertices inside the large thick circle represent the initial vertices. We use a shadowed box to abstract it into the remaining part of the Kripke structure. The right graph is constructed by connecting the initial vertices of the left and a copy with two directed edges in opposite directions, and deleting initial self-loops and one vertex from I.

We can use Lemma 3 to find a convergence law for LTL and $\mathbb{K}^1_{n,\hat{p}}$. If the LTL-formula $\sigma \to \varphi$ is not a tautology, then $\mathbb{K}^1_{n,\hat{p}}(\sigma)$ will almost surely contain a counterexample to φ and $\mathbf{P}(\mathbb{K}^1_{n,\hat{p}}(\sigma) \models \varphi) \xrightarrow{\infty} 0$, so $\mathbf{P}(\mathbb{K}^1_{n,\hat{p}}(\sigma) \models \varphi) \xrightarrow{\infty} 1$, otherwise. Together with the fact that $\mathbb{P}(\mathbb{K}^1_{n,\hat{p}} \models \sigma) = \pi(\sigma)$, this leads to the following result:

Theorem 3 (Convergence Law for LTL for One Initial Vertex). *Let φ be an LTL formula. Then*

$$\lim_{n \to \infty} \mathbf{P}(\mathbb{K}^1_{n,\hat{p}} \models \varphi) = \sum_{\substack{\sigma \in \mathcal{S}: \\ \sigma \to \varphi \text{ tautology}}} \pi(\sigma).$$

Example 4. Consider $\mathsf{F}a_1$ and $AP = \{a_1, a_2\}$. We have $\mathcal{S} = \{\neg a_1 \wedge \neg a_2, \neg a_1 \wedge a_2, a_1 \wedge \neg a_2, a_1 \wedge a_2\}$. It is easy to follow that $a_1 \wedge \neg a_2 \to \mathsf{F}a_1$ and $a_1 \wedge a_2 \to \mathsf{F}a_1$ are tautologies. So we have

$$\lim_{n \to \infty} \mathbf{P}(\mathbb{K}^1_{n,\hat{p}} \models \mathsf{F}a_1) = \mathbf{P}(\mathbb{K}^1_{n,\hat{p}} \models a_1 \wedge \neg a_2) + \mathbf{P}(\mathbb{K}^1_{n,\hat{p}} \models a_1 \wedge a_2)$$

$$= p_1(1 - p_2) + p_1 p_2$$

$$= p_1.$$

Note that, if φ is a tautology, then $\lim_{n \to \infty} \mathbf{P}(\mathbb{K}^1_{n,\hat{p}} \models \varphi) = 1$, as for $\mathbb{K}^+_{n,p}$.

4.3 Computational Complexity

By Theorem 2, determining $\mathbf{P}(\mathbb{K}_{n,\boldsymbol{p}}^{+} \models \varphi) \xrightarrow{\infty} 1$ is equivalent to determining whether φ is a tautology. For LTL formulae, this is known to be PSPACE-complete [34]. Computing $\lim_{n\to\infty} \mathbf{P}(\mathbb{K}_{n,\hat{\boldsymbol{p}}}^{1} \models \varphi)$ comes down to checking which formulae of the form $\sigma \to \varphi$ are tautologies for all $\sigma \in \mathcal{S}$. This problem is PSPACE-hard, which can be shown by considering the special case $\varphi = \mathsf{G}\psi$ and proving that computing $\lim_{n\to\infty} \mathbf{P}(\mathbb{K}_{n,\boldsymbol{p}}^{+} \models \varphi)$ is equivalent to checking whether ψ is a tautology – a problem already known to be PSPACE-hard.

Theorem 4. *Determining* $\lim_{n\to\infty} \mathbf{P}(\mathbb{K}_{n,\boldsymbol{p}}^{+} \models \varphi)$ *is PSPACE-complete, and computing* $\lim_{n\to\infty} \mathbf{P}(\mathbb{K}_{n,\hat{\boldsymbol{p}}}^{1} \models \varphi)$ *is PSPACE-hard.*

Furthermore, via Theorems 2 and 3 one can find the limit probabilities via existing LTL tautology checkers (or equivalently, satisfiability checkers), for which exponential-time algorithms exist.

5   0-1 and Convergence Laws for CTL

CTL is a temporal logic that uses atomic propositions and logical operators of LTL, along with two quantifiers, A (universal) and E (existential), to express properties over branching structures, where formulae can hold for all or some paths starting from a given vertex. Our results for LTL do not carry over to CTL: there are non-tautology CTL formulae with asymptotic probability 1, as the following example shows:

Example 5. Consider a CTL formula $\mathsf{EX}a_1$. It is not a tautology since any Kripke structure K with $L_1 = \varnothing$ has $K \not\models \mathsf{EX}a_1$. On the other hand, one can show that $\mathbf{P}(\mathbb{K}_{n,\boldsymbol{p}}^{+} \models \mathsf{EX}a_1) \xrightarrow{\infty} 1$. The idea is as follows: by Theorem 11, with limit probability 1, any induced sub-digraph isomorphism from a singleton to $\mathbb{K}_{n,\boldsymbol{p}}^{+}$ can be extended to an induced sub-digraph isomorphism $D_2 \to \mathbb{K}_{n,\boldsymbol{p}}^{+}$, where D_2 has the singleton v and another vertex v' with $v' \in L_1$; furthermore, $(v, v') \in R$. If w is the image of v in $\mathbb{K}_{n,\boldsymbol{p}}^{+}$, then this extension shows that there is a $w' \in \mathbb{K}_{n,\boldsymbol{p}}^{+}$ with $(w, w') \in R$ and $\mathbb{K}_{n,\boldsymbol{p}}^{+}, w' \models a_1$; hence $\mathbb{K}_{n,\boldsymbol{p}}^{+}, w \models \mathsf{EX}a_1$. Figure 1 shows one case of such an extension. Since with limit probability 1 we can do this for all initial w, we conclude that $\mathbf{P}(\mathbb{K}_{n,\boldsymbol{p}}^{+} \models \mathsf{EX}a_1) \xrightarrow{\infty} 1$.

CTL is a branching-time logic, meaning it models time as a tree structure. This requires us to consider all reachable vertices globally, rather than relying on a local sub-Kripke structure to draw conclusions. For instance, let K' be a Kripke structure where all vertices are in L_1. In this case, $K' \not\models \mathsf{EF}\neg a_1$. However, there exists a Kripke structure K that contains K' as an induced subgraph, where the only initial vertex can reach a vertex not in L_1. Thus, $K \models \mathsf{EF}\neg a_1$. As a result, the proof technique used to show that LTL satisfies a 0-1 law does not apply here.

This distinction makes CTL "easier" than LTL in this context. In LTL, the existence of a single counterexample path can invalidate a formula, meaning that only tautologies lack counterexamples—making satisfiability a "hard" problem. In contrast, CTL considers both the initial vertex and the reachability of various vertices. Since, in a random Kripke structure, any initial vertex almost surely has access to a wide range of states, satisfiability can almost surely be determined based on properties of the initial vertex alone.

However, CTL is a sublogic of IFPL. Since IFPL has been shown to satisfy a 0-1 law [3] for Erdős–Rényi random graphs, it follows from Proposition 1 that CTL also adheres to this law:

Theorem 5 (0-1 Law for CTL [3]). $\lim_{n\to\infty} \mathbf{P}(\mathbb{K}^+_{n,p} \models \varphi) \in \{0,1\}$ *for any CTL formula φ.*

The work [3] does not directly address computing $\lim_{n\to\infty} \mathbf{P}(\mathbb{D}^+_{n,p} \models \varphi)$ for an IFPL formula φ. However, their proof suggests an exponential-time algorithm for this computation, which can be applied to compute $\lim_{n\to\infty} \mathbf{P}(\mathbb{K}^+_{n,p} \models \varphi)$ for a CTL formula φ. In this section, we introduce a novel quadratic-time algorithm to compute $\lim_{n\to\infty} \mathbf{P}(\mathbb{K}^+_{n,p} \models \varphi)$. The algorithm of [3] is not applicable for this purpose, as it does not allow us to treat $\mathbb{D}^+_{n,p}$ and $\mathbb{K}^+_{n,p}$ as identical. In contrast, our algorithm is directly applicable to $\mathbb{K}^+_{n,p}$ and provides a novel self-contained proof of Theorem 5.

The cornerstone of our algorithm is the notion of asymptotical equivalence. For a CTL formula φ and a Kripke structure K, we let V_φ be the set of vertices $v \in V$ such that $v \models \varphi$.

Definition 6 (Asymptotical Equivalence). *Two CTL formulae φ_1, φ_2 are asymptotically equivalent (notation: $\varphi_1 \equiv_{a.e.} \varphi_2$) if $\lim_{n\to\infty} \mathbf{P}(V_{\varphi_1} = V_{\varphi_2}) = 1$.*

Note that this probability does not depend on which vertices are initial, so $\equiv_{a.e.}$ is the same for $\mathbb{K}^+_{n,p}$ and $\mathbb{K}^1_{n,\hat{p}}$. Clearly, if $\varphi \equiv_{a.e.} \varphi'$, then

$$\lim_{n\to\infty} \mathbf{P}(\mathbb{K}^+_{n,p} \models \varphi) = \lim_{n\to\infty} \mathbf{P}(\mathbb{K}^+_{n,p} \models \varphi'),$$

and likewise for $\mathbb{K}^1_{n,\hat{p}}$. Our main result is that every CTL formula is asymptotically equivalent to a propositional formula:

Theorem 6. *Let φ be a CTL formula. Then there exists a propositional formula α, independent of p, such that $\varphi \equiv_{a.e.} \alpha$. Furthermore, such α can be found in $\mathcal{O}(|\varphi|^2)$ time.*

Note that for the time complexity we again assume k fixed. Combining this with Proposition 2 then immediately gives us the following result:

Theorem 7 (0-1, Convergence, Complexity for CTL).

1. *CTL follows a 0-1 law for $\mathbb{K}^+_{n,p}$ (Theorem 5).*
2. *CTL follows a convergence law for $\mathbb{K}^1_{n,\hat{p}}$.*
3. *$\lim_{n\to\infty} \mathbf{P}(\mathbb{K}^+_{n,p} \models \varphi)$ and $\lim_{n\to\infty} \mathbf{P}(\mathbb{K}^1_{n,\hat{p}} \models \varphi)$ are found in $\mathcal{O}(|\varphi|^2)$ time.*

5.1   Idea of Proof of Theorem 6

The proof of Theorem 6 relies on two key Lemmas. The first one is that asymptotic equivalence behaves well under substitution, which we define as follows. Let φ, φ_1 and φ_2 be three CTL formulae, with φ_1 a subformula of φ. We use $\varphi\{\varphi_1 \leftarrow \varphi_2\}$ to denote the substitution of φ_2 for φ_1 in φ, which is the formula obtained by replacing φ_1 in φ with φ_2.

Lemma 4. *Let $\varphi, \varphi_1, \varphi_2$ be CTL formulae such that φ_1 is a subformula of φ and $\varphi_1 \equiv_{a.e} \varphi_2$. Then $\varphi \equiv_{a.e.} \varphi\{\varphi_1 \leftarrow \varphi_2\}$.*

The proof of this lemma relies on the fact that V_φ can be defined recursively from its subformulae; thus changing a subformula with an asymptotically equivalent one has no impact as $n \to \infty$. The second lemma gives us the tools to transform CTL formulae into propositional formulae, while staying in the same asymptotical equivalence class:

Lemma 5. *Let α_1 and α_2 be propositional formulae. Then:*

$$\mathsf{EX}\alpha_1 \equiv_{a.e.} \mathsf{EF}\alpha_1 \equiv_{a.e.} \begin{cases} false, & if\ \alpha_1 \equiv false, \\ true, & otherwise, \end{cases} \tag{1}$$

$$\mathsf{AX}\alpha_1 \equiv_{a.e.} \mathsf{AG}\alpha_1 \equiv_{a.e.} \begin{cases} true, & if\ \alpha_1 \equiv true, \\ false, & otherwise, \end{cases} \tag{2}$$

$$\mathsf{AF}\alpha_1 \equiv_{a.e.} \mathsf{EG}\alpha_1 \equiv_{a.e.} \mathsf{AU}(\alpha_2, \alpha_1) \equiv_{a.e.} \alpha_1, \tag{3}$$

$$\mathsf{EU}(\alpha_1, \alpha_2) \equiv_{a.e.} \begin{cases} false, & if\ \alpha_2 \equiv false, \\ \alpha_1 \vee \alpha_2, & otherwise. \end{cases} \tag{4}$$

Each of these statements is proven using Theorem 1; for instance, the proof for $\mathsf{EX}\alpha_1$ is sketched in Example 5.

To obtain α from φ in Theorem 6, recall that CTL is given by the following grammar:

$$\varphi ::= true \mid false \mid a_i \mid \neg\varphi \mid \varphi \wedge \varphi \mid \varphi \vee \varphi \mid \mathsf{AG}\varphi \mid$$
$$\mathsf{EG}\varphi \mid \mathsf{AF}\varphi \mid \mathsf{EF}\varphi \mid \mathsf{AX}\varphi \mid \mathsf{EX}\varphi \mid \mathsf{AU}(\varphi, \varphi) \mid \mathsf{EU}(\varphi, \varphi).$$

Now consider the syntax tree of φ: the leaves are a_i (or *true* or *false*), and intermediate vertices represent subformulae. As such, the lowest instances of temporal operators EX, EF, etc. will be in subformulae of the form $\mathsf{EX}\alpha, \mathsf{EF}\alpha$, etc. for a propositional formula α. Our algorithm works bottom-up in the syntax tree: each time we encounter such an $\mathsf{EX}\alpha, \mathsf{EF}\alpha$, we replace it by a propositional formula according to Lemma 5. By Lemma 4, this does not impact the asymptotical equivalence class of the formula as a whole. When we reach the root, we have removed all temporal operators, leaving us with a propositional formula asymptotically equivalent to φ.

Example 6. Consider $\varphi = \mathsf{AG}(\mathsf{EU}(\mathsf{AF}a_1, \neg a_1)) \wedge \mathsf{EG}(a_2)$. We repeatedly apply Lemma 5, where the blue part is the subformula being replaced.

$$\mathsf{AG}(\mathsf{EU}(\mathsf{AF}a_1, \neg a_1)) \wedge \mathsf{EG}(a_2) \equiv_{a.e.} \mathsf{AG}(\mathsf{EU}(a_1, \neg a_1)) \wedge \mathsf{EG}(a_2)$$
$$\equiv_{a.e.} \mathsf{AG}(a_1 \vee \neg a_1) \wedge \mathsf{EG}(a_2)$$
$$\equiv_{a.e.} \mathit{true} \wedge \mathsf{EG}(a_2)$$
$$\equiv_{a.e.} \mathit{true} \wedge a_2 \equiv_{a.e.} a_2.$$

Thus $\alpha = a_2$ is a valid choice for Theorem 6. It follows (see also the discussion in Sect. 3.1) that

$$\lim_{n \to \infty} \mathbf{P}(\mathbb{K}^+_{n,\boldsymbol{p}} \models \varphi) = \lim_{n \to \infty} \mathbf{P}(\mathbb{K}^+_{n,\boldsymbol{p}} \models a_2) = 0,$$
$$\lim_{n \to \infty} \mathbf{P}(\mathbb{K}^1_{n,\hat{\boldsymbol{p}}} \models \varphi) = \lim_{n \to \infty} \mathbf{P}(\mathbb{K}^1_{n,\hat{\boldsymbol{p}}} \models a_2) = p_2.$$

With regards to the time complexity, we have to traverse $\mathcal{O}(|\varphi|)$ subformulae in the syntax tree; and at every subformula, we may need to check a propositional formula to be a tautology, which as in Sect. 3 takes $\mathcal{O}(|\varphi|)$ time (recall that k is fixed). Thus α can be found in $\mathcal{O}(|\varphi|^2)$ time.

6   Conclusions and Future Work

This paper investigated the asymptotic behaviour of LTL and CTL for Erdős-Rényi-like random Kripke structures. We found that both logics follow a 0-1 law for Kripke structures with multiple initial vertices, and a convergence law for Kripke structures with a single initial vertex. We also provide algorithms to determine the limit probabilities; this has quadratic time complexities for CTL, and is PSPACE-hard for LTL.

This work is only a first step in studying model checking for random transition systems, and there are many directions to be explored. Especially interesting is the study of other random graph models such as [2], configuration [28, Chapter 13.2], and Chung-Lu [5], or Erdős-Rényi models whose probabilities vary with n [11]. The choice for other graph models should be motivated by how well they resemble real-world Kripke structures, or transition systems in general, which will require empirical investigation. It would also be interesting to see to what extent our approach could be extended to different logics, such as metric temporal logic [29] and combined logics [10].

Acknowledgements. This research has been partially funded by NWO under the grant PrimaVera (https://primavera-project.com) number NWA.1160.18.238, by the ERC Consolidator grant CAESAR number 864075, and by the European Union's Horizon 2020 research and innovation programme under the Marie Sklodowska-Curie grant agreement No. 101008233.

A   Proof of Proposition 1

A random labelled digraph

$$\mathbb{D}^{+}_{n,\boldsymbol{p}} = (V, R, I, L_1, \ldots, L_k)$$

is a Kripke structure if it has two properties:

(A) for all v there is a v' such that $(v, v') \in R$;
(B) $I \neq \varnothing$.

A random labelled digraph $\mathbb{D}^{1}_{n,\hat{\boldsymbol{p}}} = (V, R, I, L_1, \ldots, L_k)$ is a Kripke structure if it has the above property (A).

Since (A) is about R and (B) about I, these are independent, and

$$\mathbf{P}(\mathbb{D}^{+}_{n,\boldsymbol{p}} \text{ is a Kripke structure})$$
$$= \mathbf{P}(\mathbb{D}^{+}_{n,\boldsymbol{p}} \text{ satisfies (A)}) \cdot \mathbf{P}(\mathbb{D}^{+}_{n,\boldsymbol{p}} \text{ satisfies (B)}).$$

We now consider these probabilities one by one. We have

$$\mathbf{P}(\mathbb{D}^{+}_{n,\boldsymbol{p}} \text{ satisfies (B)}) = 1 - (1 - p_I)^n \xrightarrow{\infty} 1.$$

It now suffices to show that (A) holds with limit probability 1. For a fixed v, the probability of such a v' existing is $1 - (1 - p_R)^n$, so it holds for all v with probability $(1 - (1 - p_R)^n)^n$. Some calculus shows that this goes to 1 as $n \to \infty$: by L'Hôpital's rule,

$$\lim_{n \to \infty} \log\left((1 - (1 - p_R)^n)^n\right)$$
$$= \lim_{n \to \infty} n \log\left(1 - (1 - p_R)^n\right)$$
$$= \lim_{n \to \infty} \left(\frac{dn}{dn} \cdot \frac{d\log(1 - (1 - p_R)^n)}{dn}\right)$$
$$= \lim_{n \to \infty} \left(\frac{(1 - p_R)^n \log(1 - p_R)}{1 - (1 - p_R)^n}\right)$$
$$= 0,$$

so $(1 - (1 - p_R)^n)^n \xrightarrow{\infty} 1$.
Since $\mathbf{P}(\mathbb{D}^{+}_{n,\boldsymbol{p}} \text{ satisfies (A)}) = \mathbf{P}(\mathbb{D}^{1}_{n,\hat{\boldsymbol{p}}} \text{ satisfies (A)})$,

$$\lim_{n \to \infty} \mathbf{P}(\mathbb{D}^{1}_{n,\hat{\boldsymbol{p}}} \text{ is a Kripke structure})$$
$$= \lim_{n \to \infty} \mathbf{P}(\mathbb{D}^{1}_{n,\hat{\boldsymbol{p}}} \text{ satisfies (A)})$$
$$= 1.$$

B   Proof of Theorem 2

If φ is a tautology, then clearly $\lim_{n\to\infty} \mathbf{P}(\mathbb{K}_{n,p}^+ \models \varphi) = 1$. On the other hand, suppose that φ is not a tautology, and let K be a Kripke structure and π be a path in K (starting in I) such that $K, \pi \not\models \varphi$. If K' is another Kripke structure that has K as a sub-Kripke structure, then π also exists in K', so $K' \not\models \varphi$. Hence

$$\mathbf{P}(\mathbb{K}_{n,p}^+ \models \varphi) \leq 1 - \mathbf{P}(\mathbb{K}_{n,p}^+ \text{ contains } K \text{ as}$$
$$\text{a sub-Kripke structure}),$$

and the RHS goes to 0 by Lemma 1.

C   Proof of Lemma 2

Let $\mathbb{K}_{n,\hat{p}}^1(\sigma, \ominus)$ be the random Kripke structure $\mathbb{K}_{n,\hat{p}}^1$, conditioned on the properties that $\mathbb{K}_{n,\hat{p}}^1 \models \sigma$ and has no initial self-loop. Similarly, let $\mathbb{K}_{n,\hat{p}}^1(\sigma, \oplus)$ represent the random Kripke structure $\mathbb{K}_{n,\hat{p}}^1$, conditioned on $\mathbb{K}_{n,\hat{p}}^1 \models \sigma$ and the presence of the initial self-loop. We only consider $\mathbb{K}_{n,\hat{p}}^1(\sigma, \ominus)$, since if K is a sub-Kripke structure of $\mathbb{K}_{n,\hat{p}}^1(\sigma, \ominus)$, then K remains a sub-Kripke structure of $\mathbb{K}_{n,\hat{p}}^1(\sigma, \oplus)$ even when an initial self-loop is present.

In order to apply the extension statement, we need to move to labelled digraphs without initial vertices, which we formally define as follows. Recall that $\hat{p} = (p_R, p_1, \ldots, p_k)$. Define $\mathbb{D}_{n,\hat{p}} = (V, R, L_1, \ldots, L_k)$ as follows:

- V is a set of vertices with $|V| = n$.
- For each pair of vertices $v, v' \in V$, the probability that (v, v') is in the relation R is given by $\mathbf{P}((v, v') \in R) = p_R$.
- For each vertex $v \in V$ and $i = 1, \ldots, k$, the probability that $v \in L_i$ is $\mathbf{P}(v \in L_i) = p_i$.

Since $\mathbb{D}_{n,\hat{p}}$ is a $\mathscr{S}$-structure with $\mathscr{S} = \{R, L_1, \ldots, L_k\}$ with binary relation R and unary relations L_is. So, extension statement holds for $\mathbb{D}_{n,\hat{p}}$ (see [4, Theorem 6]).

We define the *random Kripke structure without initial vertices* $\mathbb{K}_{n,\hat{p}}$ by

$$\mathbf{P}(\mathbb{K}_{n,\hat{p}} = D) = \frac{\mathbf{P}(\mathbb{D}_{n,\hat{p}} = D)}{\mathbf{P}(\mathbb{D}_{n,\hat{p}} \text{ is left-total})},$$

for all labelled left-total digraphs D.

By using a similar proof to that of Proposition 1, we obtain

$$\lim_{n\to\infty} \mathbf{P}(\mathbb{D}_{n,\hat{p}} \text{ is left-total}) = 1,$$

which implies that we can treat $\mathbb{D}_{n,\hat{p}}$ and $\mathbb{K}_{n,\hat{p}}$ as identical. Let D_1 and D_2 be labelled digraphs where D_1 is a labelled sub-digraph of D_2. Since the extension statement holds for $\mathbb{D}_{n,\hat{p}}$, we state the claim:

Claim. The probability of $\mathbb{K}_{n,\hat{p}}$ having the following property goes to 1 as $n \to \infty$: Every induced sub-digraph isomorphism $f_1 \colon D_1 \to \mathbb{K}_{n,\hat{p}}$ can be extended to an induced sub-digraph isomorphism $f_2 \colon D_2 \to \mathbb{K}_{n,\hat{p}}$.

Consider a labelled digraph D_1, where a single vertex u satisfies σ $(u \models \sigma)$ and $I = R = \varnothing$, and let D_2 be obtained from K by removing the vertex from I. By the claim, any induced sub-digraph isomorphism $f_1 : D_1 \to \mathbb{K}_{n,\hat{p}}$ can be extended to an induced sub-digraph isomorphism $f_2 : D_2 \to \mathbb{K}_{n,\hat{p}}$.

Since $\mathbb{K}_{n,\hat{p}}^1(\sigma, \ominus)$ is a random Kripke structure where $v \in I$ is chosen uniformly at random from $\mathbb{K}_{n,\hat{p}}$ with $v \models \sigma$ and $(v,v) \notin R$, by setting $f'(u) = f(u) = v$ and $f'(w) = f(w)$ for all w, we can extend $f_1' : D_1' \to \mathbb{K}_{n,\hat{p}}^1(\sigma, \ominus)$ to $f_2' : D_2 \to \mathbb{K}_{n,\hat{p}}^1(\sigma, \ominus)$, where D_1' is obtained from D_1 by placing u in I and $D_2' = K$.

This implies that

$$\lim_{n \to \infty} \mathbf{P}(\mathbb{K}_{n,\hat{p}}^1(\sigma, \ominus) \text{ contains } K \text{ as an induced sub-Kripke structure}) = 1.$$

Hence, we also have

$$\lim_{n \to \infty} \mathbf{P}(\mathbb{K}_{n,\hat{p}}^1(\sigma, \ominus) \text{ contains } K \text{ as a sub-Kripke structure}) = 1.$$

D　Proof of Theorem 4

By Theorem 2, determining whether $\lim_{n \to \infty} \mathbf{P}(\mathbb{K}_{n,p}^+ \models \varphi) = 1$ is equivalent to determining whether φ is a tautology, which is a PSPACE-complete problem.

Next we consider $\mathbb{K}_{n,\hat{p}}^1$. To prove PSPACE-hardness, we focus on a special case where $\varphi = \mathsf{G}(\psi)$ with ψ being an LTL formula. If ψ is a tautology, then $\lim_{n \to \infty} \mathbf{P}(\mathbb{K}_{n,\hat{p}}^1 \models \mathsf{G}(\psi)) = 1$. Conversely, if ψ is not a tautology, then there exists a one-initial vertex Kripke structure K with the initial vertex v such that $K, v \not\models \psi$. Let D be the labelled digraph obtained from K by removing v from I. We can get the following analogue of Lemma 1 , we have

$$\lim_{n \to \infty} \mathbf{P}(\mathbb{K}_{n,\hat{p}}^1 \text{ contains } D \text{ as a labelled sub-digraph}) = 1.$$

Consequently, we have

$$\lim_{n \to \infty} \mathbf{P}(\text{there exits a vertex in } \mathbb{K}_{n,\hat{p}}^1 \text{ such that } v \not\models \psi) = 1.$$

Additionally, it follows from (Theorem 11.9 [11])

$$\lim_{n \to \infty} \mathbf{P}(\mathbb{K}_{n,\hat{p}}^1 \text{ is strongly connected}) = 1$$

that $\lim_{n \to \infty} \mathbf{P}(\text{there exists a path from the initial vertex to } v \text{ in } \mathbb{K}_{n,\hat{p}}^1) = 1$. Combining these results, we conclude that

$$\lim_{n \to \infty} \mathbf{P}(\text{there exists a path stating from the initial vertex in } \mathbb{K}_{n,\hat{p}}^1$$
$$\text{not satisfying } \mathsf{G}(\psi)) = 1.$$

So we have if ψ is not a tautology, then $\lim_{n\to\infty} \mathbf{P}(\mathbb{K}^1_{n,\hat{p}} \models \mathsf{G}\psi) = 0$. Thus, this problem for $\varphi = \mathsf{G}(\psi)$ is equivalent to determining whether ψ is a tautology. Since checking if ψ is a tautology is PSPACE-hard, it follows that this problem is PSPACE-hard.

E   Proof of Lemma 5

As previously mentioned, we treat $\mathbb{K}^+_{n,p}$ and $\mathbb{K}^1_{n,\hat{p}}$ as equivalent. Therefore, in the following proof, we focus solely on $\mathbb{K}^+_{n,p}$.

(1) We only give the proof of $\mathsf{EX}\alpha_1$ since $\mathsf{EF}\alpha_1$ is similar. If $\alpha_1 \equiv \mathit{false}$, the result holds since $V_{\mathsf{EX}\mathit{false}} = \varnothing = V_{\mathit{false}}$. Now suppose $\alpha_1 \not\equiv \mathit{false}$. Let C be the set of propositional formulae of the form $l_1 \wedge l_2 \wedge \ldots \wedge l_k$, where every l_i is equal to either a_i or $\neg a_i$. For $\alpha \in C$, let D_2 be a labelled digraph with $V = \{v_1, v_2\}$, $R = (v_1, v_2)$, and choose the L_i such that $v_1 \in I$, and such that v_1 satisfies α and v_2 satisfies α_1. Note that α determines whether $v_1 \in L_i$ or $v_1 \notin L_i$. Let D_1 be the sub-digraph with vertex set $\{v_1\}$. Applying Theorem 1 to (D_1, D_2) shows that with limit probability 1, $\mathbb{K}^+_{n,p}, v \models \mathsf{EX}\alpha_1$ for all $v \in V_\alpha$. Since $\bigcup_{\alpha \in C} V_\alpha = V$, this proves the equation.
(2) We only give the proof of $\mathsf{AX}\alpha_1$ since $\mathsf{AF}\alpha_1$ is similar. If $\alpha \equiv \mathit{true}$, the result holds since $V_{\mathsf{AX}\mathit{true}} = V = V_{\mathit{true}}$. If not, then $\neg\mathsf{AX}\alpha_1 \equiv \mathsf{EX}\neg\alpha_1$, and $\lim_{n\to\infty} \mathbf{P}(V_{\mathsf{AX}\varphi_1} = \varnothing = V_{\mathit{false}}) = 1$ by (1).
(3) We only give the proof of $\mathsf{AF}\alpha_1$ and $\mathsf{AU}(\alpha_2, \alpha_1)$ since $\mathsf{EG}\alpha_1$ is similar. We start with $\mathsf{AF}\alpha_1$. If $\mathbb{K}_{n,p}, v \models \alpha_1$, then $\mathbb{K}_{n,p}, v \models \mathsf{AF}\alpha_1$. Therefore, $V_{\alpha_1} \subseteq V_{\mathsf{AF}\alpha_1}$. We now show $\mathbf{P}(V_{\mathsf{AF}\alpha_1} \subseteq V_{\alpha_1}) \xrightarrow{\infty} 1$ by showing $\mathbf{P}(V_{\neg\alpha_1} \subseteq V_{\neg\mathsf{AF}\alpha_1}) \xrightarrow{\infty} 1$: Applying Theorem 1 with D_1 a vertex satisfying $\neg\alpha_1$ and D_2 a directed cycle with all vertices satisfying $\neg\alpha_1$ shows that with limit probability 1, each vertex v not satisfying $\neg\alpha_1$ lies on a cycle of vertices satisfying $\neg\alpha_1$; thus $v \in V_{\neg\mathsf{AF}\alpha_1}$. For $\mathsf{AU}(\alpha_2, \alpha_1)$, it follows from $V_{\alpha_1} \subseteq V_{\mathsf{AU}(\alpha_2,\alpha_1)} \subseteq V_{\mathsf{AF}\alpha_1}$ that $\mathbf{P}(V_{\mathsf{AU}(\alpha_2,\alpha_1)} = V_{\alpha_1}) \xrightarrow{\infty} 1$.
(4) If $\alpha_2 \equiv \mathit{false}$, then $V_{\mathsf{EU}(\alpha_1,\mathit{false})} = V_{\mathit{false}} = V_{\alpha_2} = \varnothing$.
Otherwise note that $V_{\alpha_2} \cup (V_{\alpha_1} \cap V_{\mathsf{EX}\alpha_2}) \subseteq V_{\mathsf{EU}(\alpha_1,\alpha_2)} \subseteq V_{\alpha_1 \vee \alpha_2}$. By (1), $\mathbf{P}(V_{\mathsf{EX}\alpha_2} = V) \xrightarrow{\infty} 1$, so $\mathbf{P}(V_{\alpha_2} \cup (V_{\alpha_1} \cap V_{\mathsf{EX}\alpha_2}) = V_{\alpha_1} \cup V_{\alpha_2}) \xrightarrow{\infty} 1$. Since $V_{\alpha_1} \cup V_{\alpha_2} = V_{\alpha_1 \vee \alpha_2}$ this proves the equation.

References

1. Baier, C., Katoen, J.P.: Principles of Model Checking. The MIT Press, Cambridge (2008)
2. Barabási, A.L., Albert, R.: Emergence of scaling in random networks. Science **286**, 509–512 (1999)
3. Blass, A., Gurevich, Y., Kozen, D.: A zero-one law for logic with a fixed-point operator. Inf. Control **67**(1–3), 70–90 (1985)
4. Blass, A., Harary, F.: Properties of almost all graphs and complexes. J. Graph Theor. **3**(3), 225–240 (1979)

5. Chung, F., Lu, L.: The average distances in random graphs with given expected degrees. Proc. Natl. Acad. Sci. **99**(25), 15879–15882 (2002)
6. Compton, K.: 0-1 laws in logic and combinatorics. In: Rival, I. (ed.) Proceedings of the NATO Advanced Study Institute on Algorithms and Order, pp. 1–29. D. Reidel, Dordrecht (1988)
7. Dawar, A., Grädel, E.: Properties of almost all graphs and generalized quantifiers. Fund. Inform. **98**(4), 351–372 (2010)
8. Erdös, P., Rényi, A.: On random graphs I. Publicationes Mathematicae Debrecen **6**, 290–297 (1959)
9. Fagin, R.: Probabilities on finite models. J. Symbolic Logic **41**(1), 50–58 (1976)
10. Franceschet, M., Montanari, A., Rijke, M.: Model checking for combined logics with an application to mobile systems. Autom. Softw. Eng. **11**, 289–321 (2004)
11. Frieze, A., Karoński, M.: Introduction to Random Graphs. Cambridge University Press, Cambridge (2016)
12. Frieze, A., McDiarmid, C.: Algorithmic theory of random graphs. Random Struct. Algorithms **10**(1–2), 5–42 (1997)
13. Glebskii, Y.V., Kogan, D.I., Liogon'kii, M.I., Talanov, V.A.: Range and degree of realizability of formulas in the restricted predicate calculus. Cybern. Syst. Analysis, Translated Kibernetika **5**(2), 142–154 (1969)
14. Grandjean, E.: Complexity of the first-order theory of almost all finite structures. Inf. Control **57**(3), 180–204 (1983)
15. Haber, S., Hershko, T., Mirabi, M., Shelah, S.: First-order logic with equicardinality in random graphs. In: Endrullis, J., Schmitz, S. (eds.) 33rd EACSL Annual Conference on Computer Science Logic (CSL 2025). Leibniz International Proceedings in Informatics (LIPIcs), vol. 326, pp. 12:1–12:17. Schloss Dagstuhl – Leibniz-Zentrum für Informatik, Dagstuhl, Germany (2025)
16. van Hee, K.M., Liu, Z.: Generating benchmarks by random stepwise refinement of Petri nets. In: Donatelli, S., Kleijn, J., Machado, R.J., Fernandes, J.M. (eds.) Proceedings of the Workshops of the 31st International Conference on Application and Theory of Petri Nets and Other Models of Concurrency (PETRI NETS 2010) and of the 10th International Conference on Application of Concurrency to System Design (ACSD 2010), Braga, Portugal, June, 2010. CEUR Workshop Proceedings, vol. 827, pp. 403–417. CEUR-WS.org (2010)
17. Heinig, P., Müller, T., Noy, M., Taraz, A.: Logical limit laws for minor-closed classes of graphs. J. Comb. Theory Ser. B **130**, 158–206 (2018)
18. Huisman, M., Wijs, A.: Concise Guide to Software Verification - From Model Checking to Annotation Checking, 2. Texts in Computer Science, Springer (2023)
19. Hussain, I., Csallner, C., Grechanik, M., Fu, C., Xie, Q., Park, S., Taneja, K., Hossain, B.M.M.: Evaluating program analysis and testing tools with the rugrat random benchmark application generator. In: Proceedings of the Ninth International Workshop on Dynamic Analysis, pp. 1–6. WODA 2012, Association for Computing Machinery, New York, NY, USA (2012)
20. Kaufmann, M.: A counterexample to the 0-1 law for existential monadic second-order logic. Computational Logic Inc, pp. 1–5 (1988)
21. Kaufmann, M., Shelah, S.: On random models of finite power and monadic logic. Discret. Math. **54**(3), 285–293 (1985)
22. Khoussainov, B., Nerode, A.: Automata Theory and Its Applications. Birkhauser Boston Inc, USA (2001)
23. Kolaitis, P.G., Vardi, M.Y.: The decision problem for the probabilities of higher-order properties. In: Proceedings of the Nineteenth Annual ACM Symposium on

Theory of Computing, pp. 425–435. STOC '87, Association for Computing Machinery, New York (1987)

24. Kolaitis, P.G., Vardi, M.Y.: Infinitary logics and 0–1 laws. Inf. Comput. **98**(2), 258–294 (1992)
25. Lynch, J.F.: Probabilities of sentences about very sparse random graphs. Random Struct. Algorithms **3**(1), 33–53 (1992)
26. Malyshkin, Y.A., Zhukovskii, M.E.: MSO 0–1 law for recursive random trees. Stat. Probab. Lett. **173**, 109061 (2021)
27. Mealy, G.H.: A method for synthesizing sequential circuits. Bell Syst. Tech. J. **34**(5), 1045–1079 (1955)
28. Newman, M.E.: Networks: An Introduction. Oxford University Press, Oxford (2010)
29. Ouaknine, J., Worrell, J.: On the decidability of metric temporal logic. In: 20th Annual IEEE Symposium on Logic in Computer Science, pp. 188–197. LICS' 05, IEEE, Chicago (2005)
30. Rabin, M., Scott, D.: Finite automata and their decision problems. IBM J. Res. Dev. **3**, 114–125 (1959)
31. Razafimahatratra, A.S., Zhukovskii, M.E.: Zero–one laws for k-variable first-order logic of sparse random graphs. Discrete Appl. Math. **276**, 121–128 (2020), 2nd Russian–Hungarian Combinatorial Workshop
32. Scheidgen, M.: Generation of large random models for benchmarking. In: Kolovos, D.S., Ruscio, D.D., Matragkas, N.D., Cuadrado, J.S., Ráth, I., Tisi, M. (eds.) Proceedings of the 3rd Workshop on Scalable Model Driven Engineering part of the Software Technologies: Applications and Foundations (STAF 2015) Federation of Conferences, L'Aquila, Italy, July 23, 2015. CEUR Workshop Proceedings, vol. 1406, pp. 1–10. CEUR-WS.org (2015)
33. Shelah, S.: On failure of 0–1 laws. In: Beklemishev, L.D., Blass, A., Dershowitz, N., Finkbeiner, B., Schulte, W. (eds.) Fields of Logic and Computation II: Essays Dedicated to Yuri Gurevich on the Occasion of His 75th Birthday, pp. 293–296. Springer, Cham (2015)
34. Sistla, A.P., Clarke, E.M.: The complexity of propositional linear temporal logics. J. ACM **32**(3), 733–749 (1985)
35. Spencer, J.: Zero-one laws with variable probability. J. Symb. Log. **58**(1), 1–14 (1993)
36. Spencer, J.: The Strange Logic of Random Graphs. Springer, New York (2001)
37. Talanov, V.: Asymptotic solvability of logical formulas. Comb.-Algebraic Methods Appl. Math., 118–126 (1981)
38. Talanov, V., Knyazev, V.: The asymptotic truth of infinite formulas. In: Proceedings of All-Union Seminar on Discrete and Applied Mathematics and Its Applications, pp. 56–61 (1986)
39. TOOLS2009 - International Workshop on Graph-Based Tools (GraBaTs'2009): Satellite workshop to TOOLS 2009 (2009). http://is.tm.tue.nl/staff/pvgorp/events/grabats2009/

Accelerating CAR-Based Model-Checking with Multiple Unsatisfiable Cores

Yibo Dong[1], Xiwei Wu[2], Jianwen Li[1(✉)], Geguang Pu[1], and Ofer Strichman[3]

[1] Software Engineering Institute, East China Normal University, Shanghai, China
jwli@sei.ecnu.edu.cn
[2] Department of Computer Science and Engineering, Shanghai Jiao Tong University,
Shanghai, China
[3] Faculty of Data and Decision Sciences, Technion, Haifa, Israel

Abstract. Model checking is a framework for automated formal verification of transition systems, like hardware designs. Two leading model-checking techniques, PDR and CAR, are based on multiple calls to a SAT solver, for the purpose of finding a path to a bug or proving its absence. They build sequences of formulas, called *frames*, that represent approximated sets of states, e.g., a frame can contain an over-approximation of the states that can reach the negated property within a given number of steps. Part of the process is to make these sets more precise, i.e., less approximating, and this is done by strengthening the frames with negation of states that were proven to be spurious. A key component for performance is the ability of these engines to generalize those states and thus accelerate the process of making the frames more precise. This is done by strengthening them with the *unsatisfiable cores* that the SAT solver returns. In this work, we suggest several performance improvements to this process, most notably a technique for generating multiple such cores in linear time and adding them simultaneously to the frames. Our results show that our implementation of these techniques, on top of SIMPLECAR, not only improves its performance, but also solves more (unsafe) cases than any other model-checker in the public domain, and solves instances from the HWMCC that no other model checker can solve, hence it contributes to the state-of-the-art.

1 Introduction

Model checking is an automatic formal-verification technique that is central in the hardware design community [3,15]. Given a model M and a temporal property P over its variables, it checks whether all the behaviors of M satisfy P, i.e., whether $M \models P$. Once a system behavior is detected to violate P, the model checker returns a *counterexample* as the evidence, which demonstrates the execution of the system leading to the property violation. Such a process is

We thank the anonymous reviewers for their insightful feedback. This work is supported by NSFC Grant #62372178 and #U21B2015.

G. Ernst and K. Y. Rozier (Eds.): SPIN 2025, LNCS 15945, pp. 88–105, 2026.
https://doi.org/10.1007/978-3-032-06847-7_5

called *bug-finding*. If P is a *safety* property, the violation of P is witnessed by a counterexample made of a finite number of states. It is well known that model checking on safety properties can be reduced to reachability analysis [8].

State-of-the-art safety model checking techniques include Bounded Model Checking (BMC) [4,6], Interpolation Model Checking (IMC) [18], Property Directed Reachability (PDR) (also called IC3) [7,11], and Complementary Approximate Reachability (CAR) [17], all of which integrate a SAT solver internally. BMC is an incomplete method (it is only used for finding bugs, not proving their absence) and as such, is empirically very fast at finding relatively shallow bugs (i.e., after a relatively small number of steps from the initial state). IMC, PDR, and CAR are complete but are generally not as fast as BMC in shallow bug-finding, and none of the existing implementations of those techniques dominate the other. In [16,17] it was empirically shown that within a given time and hardware resources, CAR is able to solve unsafe (i.e., instances in which the property fails) instances that BMC cannot, and safety instances that IMC and PDR cannot, while the converse is true as well. Therefore, a portfolio consisting of different techniques is often maintained for different verification tasks. However, hardware model-checking (a Pspace problem) always falls short of the performance needs in the industry when it comes to verifying large designs. Indeed, performance optimization of SAT-based model checkers is an active research area. Some recent examples are [10,27] and [22].

In this paper, we focus on improving the performance of CAR. We will describe in detail how CAR works in Sect. 3. It has many similarities to PDR, which is better known, but also several distinctive features. For now, let us mention that similar to PDR, it relies on many SAT calls over relatively easy formulas. One of its elements is a sequence of formulas $O_1 \ldots O_k$, called the over-approximating frames (or *O-frames*, for short), where O_i, $1 \leq i \leq k$ over-approximates the states that can reach $\neg P$ within i steps. CAR gradually makes these frames more precise, i.e., less over-approximating, by removing from them states that cannot reach $\neg P$ within the given number of steps. One of the key elements of this process is *generalization*, that is, the ability to remove many such states at once. This is done by finding the *unsatisfiable core* (UC) of unsatisfiable SAT calls. The research that we report here is focused on finding multiple such UCs, which accelerates the narrowing process of the O-frames. To explain our contribution, let us first describe briefly how modern SAT solvers find UCs.

The input to every SAT call in CAR (and PDR) takes the form of $\bigwedge_{l \in \mathcal{A}} l \wedge \phi$, where ϕ is a Boolean formula in Conjunctive Normal Form (CNF) and $\mathcal{A}$ consists of a sequence of literals, called the *assumptions*. Almost all modern CDCL-based SAT solvers as of MINISAT [12] support assumptions. They position the literals in $\mathcal{A}$, in order, as their first decisions, and perform Boolean Constraint Propagation (BCP) as usual. Unsatisfiability is detected when the BCP of an assumption contradicts the value of another literal (because recall, all the literals in $\mathcal{A}$ and those that are implied by them via BCP are implied by the formula regardless of any decision). By analyzing the trail, the solver can detect which of the assumptions contributed to the conflict and emit this list of assumptions as the UC, which is essentially a compact *reason* for the unsatisfiability. In other words, the UC is a subset of $\mathcal{A}$ that is sufficient for making ϕ unsatisfiable.

There can be multiple UCs in a given unsatisfiable formula, and the order of the assumptions may affect the UC that is found (and this, in turn, affects the overall performance of the model-checker, whether it is CAR or PDR). More explicitly, the literals that are propagated earlier are more likely to appear in the returned UC. Indeed, prior work leveraged this phenomenon to improve performance. Specifically, the IC3ref model checker [14], which implements the original IC3 algorithm, sorts the literals in $\mathcal{A}$ in descending order based on their appearance frequencies. The SIMPLECAR model-checker, which implements CAR, uses two different literal-ordering strategies, as reported in [10]: *Intersection*, which prioritizes literals that are both in the current state and the latest generated UC, and *Rotation*, which prioritizes literals that are present in all previously explored states. This makes these literals more likely to appear as part of the generated UC and empirically improves the performance of bug-finding. Indeed, SIMPLECAR is one of the baseline implementations against which we compare our contributions.

The research that we report here, is focused on computing and adding more than one UC at a time, hence accelerating the narrowing of the O-frames. We prove that once the SAT solver proved that the formula is unsatisfiable, it is a *linear-time* operation to find multiple UCs, simply by rerunning the solver incrementally, with a different assumptions order. Since all the learned clauses remain from the initial run, it is guaranteed that the solver will detect unsatisfiability solely based on deciding the assumptions (or a subset thereof) and applying BCP. However, our experiments with variants of this approach demonstrated the difference between asymptotic complexity and actual run-time: although this linear-time operation is supposed to improve the search and thus lead to less exponential-time SAT solving, in practice the formulas given to the SAT solver in CAR are so easy that they are solved in fractions of a second. Thus, the trade-off between more cores and fewer SAT calls in practice is not at all an obvious win. We will describe several algorithmic steps that we took in order to make this technique cost-effective. Using this combination of techniques, we not only improved the average performance of SIMPLECAR but also reached the point that it solves more unsafe cases from the HWMCC15 + HWMCC17 benchmark sets than any other model checker, and furthermore it can solve several cases that have never been solved before by any model checker, thus contributing to the state-of-the-art.[1]

We continue with preliminaries in the next section. In Sect. 3, we recall the CAR model checking algorithm. In Sect. 4, we present our new methods in detail, including experiments, and in Sect. 5, we compare our results to other tools. We conclude and suggest topics for future research in Sect. 6.

---

[1] CAR with the improvements reported here has recently won the 3rd place in bit-level model-checking competition(safe + unsafe), and 2nd place in unsafe cases [13].

2   Preliminaries

2.1   Boolean Transition System

A Boolean transition system Sys is a tuple (V, I, T), where V and V' denote the set of variables in the present state and the next state, respectively. The state space of Sys is the set of possible variable assignments. I is a Boolean formula corresponding to the set of initial states, and T is a Boolean formula over $V \cup V'$, representing the transition relation. State s_2 is a successor of state s_1 iff $s_1 \wedge s_2' \models T$, which is also denoted by $(s_1, s_2) \in T$. A $path$ of length k is a finite state sequence $s_1, s_2, \ldots, s_k$, where $(s_i, s_{i+1}) \in T$ holds for $(1 \leq i \leq k - 1)$. A state t is reachable from s in k steps if there is a path of length k from s to t. Let $X \subseteq 2^V$ be a set of states in Sys. We denote the set of successors of states in X as $R(X) = \{t \mid (s, t) \in T, s \in X\}$. Conversely, we define the set of predecessors of states in X as $R^{-1}(X) = \{s \mid (s, t) \in T, t \in X\}$. Recursively, we define $R^0(X) = X$ and $R^i(X) = R(R^{i-1}(X))$ where $i > 0$, and the notation $R^{-i}(X)$ is defined analogously. In short, $R^i(X)$ denotes the states that are reachable from X in i steps, and $R^{-i}(X)$ denotes the states that can reach X in i steps.

2.2   Safety Model Checking and Reachability Analysis

Given a transition system $Sys = (V, I, T)$ and a safety property P, which is a Boolean formula over V, a model checker either proves that P holds for any state reachable from an initial state in I or disproves P by producing a *counterexample*. In the former case, we say that the system is safe, while in the latter case, it is unsafe. A counterexample is a finite path from an initial state s to a state t violating P, i.e., $t \models \neg P$, and such a state is called a *bad* state.

In symbolic model checking, safety checking is reduced to symbolic reachability analysis. Reachability analysis can be performed in a forward or backward search. Forward search starts from initial states I and searches for reachable states of I by computing $R^i(X)$ with increasing values of i, while backward search begins with states in $\neg P$ and computes $R^{-i}(X)$ with increasing values of i to search for states reaching I. Table 1 gives the corresponding formal definitions.

Table 1. Standard Reachability Analysis

	Forward	Backward
Base	$F_0 = I$	$B_0 = \neg P$
Induction	$F_{i+1} = R(F_i)$	$B_{i+1} = R^{-1}(B_i)$
Safe Check	$F_{i+1} \subseteq \bigcup_{0 \leq j \leq i} F_j$	$B_{i+1} \subseteq \bigcup_{0 \leq j \leq i} B_j$
Unsafe Check	$F_i \cap \neg P \neq \emptyset$	$B_i \cap I \neq \emptyset$

For forward search, F_i denotes the set of states that are reachable from I within i steps, which is computed by iteratively applying R. At each iteration, we first compute a new F_i and then perform safe and unsafe checking. If the condition in the safe/unsafe checking is satisfied, the search process terminates. Intuitively, if the unsafe checking passes, then some bad state is reachable, and if the safe checking passes, then all the reachable states from I have been checked, and none of them can reach a bad state. For backward search, the set B_i is the set of states that can reach $\neg P$ in i steps, and the search procedure is analogous to the forward one.

2.3 SAT Solving and Unsatisfiable Cores

In propositional logic, a *literal* is an atomic variable or its negation. A *cube* is a conjunction of literals, and a clause is a disjunction of literals. The negation of a clause is a cube, and vice versa. A formula in *Conjunctive Normal Form* (CNF) is a conjunction of clauses. For simplicity, we also treat a CNF formula ϕ as a set of clauses. Similarly, a cube or a clause c can be treated as a set of literals or a Boolean formula, depending on the context.

We say that a CNF formula ϕ is satisfiable if there exists an assignment of each Boolean variable in ϕ such that ϕ is true; otherwise, ϕ is unsatisfiable. Generally, it is an NP-complete problem to decide whether a given CNF formula is satisfiable. A SAT solver can decide whether a CNF formula ϕ is satisfiable or not. It emits a Boolean assignment to the variables, called a model of ϕ, if ϕ is satisfiable. Otherwise, some SAT solvers can emit an unsatisfiable core, as explained in the introduction, based on a subset of the assumptions.

3 Complementary Approximate Reachability (CAR)

CAR is a relatively new SAT-based safety model checking approach that is essentially a reachability-analysis algorithm inspired by PDR [17]. Unlike BMC [4,6], CAR is complete, i.e., it can also prove correctness. CAR maintains two sequences of state sets (also called 'frames') that are defined as follows:

Definition 1 (Approximating State Sequences). *Given a transition system* $Sys = (V, I, T)$ *and a safety property* P, *the over-approximating state sequence* $O \equiv O_0, O_1, \ldots, O_i$ $(i \geq 0)$, *and the under-approximating state sequence* $U \equiv U_0, U_1, \ldots, U_j$ $(j \geq 0)$ *are finite sequences of state sets such that, for* $k \geq 0$:

	O-sequence	U-sequence
Base:	$O_0 = \neg P$	$U_0 = I$
Induction:	$O_{k+1} \supseteq R^{-1}(O_k)$	$U_{k+1} \subseteq R(U_k)$
Constraint:	$O_k \cap I = \emptyset$	—

These sequences determine the termination of CAR as follows:

- *Return 'Unsafe' if $\exists i \cdot U_i \cap \neg P \neq \emptyset$.*
- *Return 'Safe' if $\exists i \geq 1 \cdot (\bigcup_{j=0}^{i} O_j) \supseteq O_{i+1}$.*

Notably, CAR can also use the over- and under- approximating sequences reversed, i.e., use the over-approximating sequence in the forward direction, from the initial state towards the negated property, while using the under-approximating sequence from the negated property towards the initial state. In this paper, we only consider the direction as stated in Definition 1 (this was called 'backward CAR' in [16,17]).

At the high level, CAR can be considered a general version of PDR, as the O-sequence in CAR is not necessarily monotone, while that in PDR is. As a result, CAR can have a more flexible methodology for the *state generalization*, i.e., directly using the UC from the SAT solver rather than computing the *relative inductive clauses*. However, CAR needs to invoke additional SAT queries to find the invariant (checking safety), while PDR can do it with a simple syntactic check.

Algorithm 1 describes CAR. It progresses by widening the U sets and narrowing the O sets, which are initialized at Line 2 to I and $\neg P$, respectively. The algorithm maintains a stack of pairs $\langle state, level \rangle$ where $level$ refers to an index of an O frame. O_{tmp}, initialized to $\neg I$ in Line 4 and later updated, represents the next frame to be created.

Initially, a state from the U-sequence is heuristically picked (Line 5) – by default from the end to the beginning – and pushed to the stack. In each iteration of the internal loop, CAR checks whether the state at the top of the stack, call it s, can transit to the O_l frame. This involves two steps. It first conducts a 'blockedIn' check (Line 11, to be discussed later) to check if the state is blocked at level $l + 1$, i.e., whether $O_{l+1} \rightarrow \neg s$. If yes, backtrack is initiated; otherwise, CAR checks if $SAT(s, T \wedge O_l')$ in Line 18, i.e., whether $T \wedge O_l'$ is satisfiable while taking the literals in s as assumptions (recall from Sect. 2.1 that prime variables, such as O_l' here, denote next-state variables).

If yes, a new state $t \in O_l$ is extracted from the model, to update the U-sequence (Line 19-21), effectively *widening* it; Alternatively, the negation of the unsatisfiable core is used to constrain the O frame of s (level $l + 1$), effectively *narrowing* it (Lines 23-30), and pushing s back to the stack if possible.

CAR returns 'Unsafe' as soon as the working level l is less than 0, which indicates that a bad state in $\neg P$ is reached (line 10). Otherwise, CAR returns 'Safe' if the O sequence includes all the states that can reach $\neg P$ – this is checked via the condition in Line 31, which was also mentioned as part of Definition 1.

Let us go back to '$blockedIn(s, l)$'. It can be thought of as an optimization – a linear-time test that saves calls to a SAT solver in line 18. It returns true if there exists a clause $cl \in O_l$ that subsumes (and hence implies) $\neg s$. Although each pair (s, l) should adhere to the condition that s is not blocked at level $l + 1$ when the pair is pushed into the stack (line 21), the successors of s might later backtrack to a higher level, giving rise to a UC that blocks s. In the algorithm, if s is already blocked in O_{l+1}, then according to the definition, it cannot reach O_l within one step, and there is no need for the SAT call in line 18.

Algorithm 1: Complementary Approximate Reachability (CAR).

Input: A transition system $Sys = (V, I, T)$ and a safety property P
Output: 'Safe' or ('Unsafe' + a counterexample)

1   **if** $SAT(I \wedge \neg P)$ **then return** 'Unsafe'
2   $U_0 := I$, $O_0 := \neg P$
3   **while** *true* **do**
4      $O_{tmp} := \neg I$
5   **while** $state := pickState(U)$ *is successful* **do** $\triangleright$ Heuristic choice
6         $stack := \emptyset$
7         $stack.push(state, |O| - 1)$
8   **while** $|stack| \neq 0$ **do**
9            $(s, l) := stack.top()$ $\triangleright$ Assume $s \in U_j$
10   **if** $l < 0$ **then return** 'Unsafe'
11   **if** $blockedIn(s, O_{l+1})$ **then** $\triangleright$ true if $O_{l+1} \to \neg s$
12              $stack.pop()$ $\triangleright$ note line 9
13   **while** $l + 1 < |O|$ **and** $blockedIn(s, O_{l+1})$ **do**
14                $l := l + 1$ $\triangleright$ Backtrack to a higher level
15   **if** $l + 1 < |O|$ **then**
16                $stack.push(s, l + 1)$
17   **continue**
18   **if** $SAT(s, T \wedge O'_l)$ **then**
19              $t := GetModel()$
20              $U_{j+1} := U_{j+1} \vee t$ $\triangleright$ Widening U. j is s's frame – see Line 9
21              $stack.push(t, l - 1)$
22   **else**
23              $stack.pop()$
24              $uc := getUC()$
25   **if** $l + 1 < |O|$ **then**
26                $O_{l+1} := O_{l+1} \wedge (\neg uc)$
27                $l' = minNotBlockedIn(s)$
28                $stack.push(s, l' - 1)$
29   **else**
30                $O_{tmp} := O_{tmp} \wedge (\neg uc)$
31   **if** $\exists i \geq 1$ $s.t.$ $(\bigcup_{0 \leq j \leq i} O_j) \supseteq O_{i+1}$ **then return** 'Safe'
32   Add a new state-set to O and initialize it to O_{tmp}

---

4   Adding Multiple Unsatisfiable Cores

4.1   Theory and Initial Results

Recall that in CAR, the O frames are narrowed (i.e., become less overapproximating) by constraining them with negations of UCs—see line 30 in Algorithm 1. We now suggest a method by which each time the formula in line 18 turns out to be unsatisfiable, more than one UC is computed and added. Constraining further the frames in this way accelerates the narrowing of the O-frames and hence blocks more spurious states from being explored. It is guaranteed, by construction, that each new core is *not* implied by previous (negations of) UCs in the frame[2], as stated in the following theorem.

Theorem 1 (Additional UCs are not redundant). *Let (s, l) be a pair checked in line 9 of Algorithm 1. If s is not blocked in O_{l+1} and $SAT(s, T \wedge O'_l)$ returns 'unsatisfiable', the UC retrieved from the SAT call, call it uc, satisfies $O_{l+1} \not\rightarrow \neg uc$.*

Proof. uc is a core of $SAT(s, T \wedge O'_l)$, hence $uc \subseteq s$, or, equivalently, $\neg uc \rightarrow \neg s$. Hence if $O_{l+1} \rightarrow \neg uc$, then $O_{l+1} \rightarrow \neg s$, which implies that s has already been blocked in O_{l+1} before the SAT call. A contradiction. $\square$

We observe that once the SAT solver proved that the formula is unsatisfiable, it is a *linear-time* operation to find multiple UCs simply by rerunning the solver incrementally, with a different assumptions order. To be specific, since all the learned clauses remain from the initial run, it is guaranteed that the solver will detect unsatisfiability solely based on deciding the assumptions (or a subset thereof) and applying BCP. More formally:

Theorem 2 (The complexity of computing additional UCs is linear). *Let $\mathcal{A}$ be the assumptions (a vector of literals) and f be a formula. In a modern CDCL SAT Solver, if $SAT(f, \mathcal{A})$ is unsatisfiable, then a subsequent incremental call $SAT(f, reorder(\mathcal{A}))$, where reorder is some reordering of the assumptions, takes time which is linear in the number of f literals.*

Proof. In a Minisat-like SAT solver, the assumptions are selected as decision literals in the order in which the assumptions vector is given to the solver, followed each time by BCP. A formula is declared unsat when a conflict occurs in some decision level d such that $d \leq |\mathcal{A}|$. Let $uc \subseteq \mathcal{A}$ denote the UC of the original SAT call. In consequent incremental runs of the solver (recall that the checked formula includes all the learned clauses from the previous run that were still present, i.e., not deleted, at the time that the conflict was detected) with any order of assumptions, a conflict must occur at or before the point in which this subset is 'covered', hence before any real decision is made. Therefore, only decisions and BCP are applied, which is linear in the number of literals. $\square$

Theorem 2 gives us hope that finding multiple UCs in a single run will be cost-effective. We call this method mUC, where m stands for multiple. Generally, we do not need to commit to a constant number of UCs, as we can heuristically decide how many to add. For exposition purposes, however, we will denote by mUC (n) a process of adding exactly n cores. mUC (1) is, therefore, the baseline of adding a single core.

---

[2] However, it is possible that the additional cores are subsumed by previous ones in the same formula. Indeed, we experimented with applying a subsumption test, but it turned out not to be cost-effective.

There is a large space for tuning this approach, including a decision on which order of assumptions to use in order to produce more UCs, how many UCs to add, whether always to apply it or just selectively, and so forth. We drove this tuning process with the 337 unsafe benchmarks[3] in the Aiger [5] format from the single safety property track of the 2015 and 2017 Hardware Model Checking Competition (HWMCC).[4] All the counterexamples found were successfully verified with the third-party tool AIGSIM that comes with the AIGER package.

Our initial experiments with variants of this approach demonstrated the difference between asymptotic complexity and actual run-time. Although this linear-time operation is supposed to improve the search and thus lead to fewer exponential-time SAT calls, in practice, the formulas given to the SAT solver in CAR are so easy that they are solved in fractions of a second. Thus, it is not obvious at all that more cores and fewer SAT calls in practice improve the total running time.

To test the actual difference in running time, we changed the algorithm. Each time the result of the SAT call was 'unsat', we ran it once more, incrementally, with the order of the literals reversed. This is not only cheap to compute, but also guarantees that the literals that have been picked in the previous run are in the back, which promotes diversity of the cores. We shall later show that this is a good choice empirically. Recall that this second run is linear according to Theorem 2, while the first run is worst-case exponential. Across all benchmarks, the ratio between the run times of the second and first runs turned out to be 0.65. In other words, if we always apply it, then each time the result of the SAT call is 'unsat', we pay a penalty of 65% in running time, with the hope that this will improve the search and consequently reduce the overall number of SAT calls and the total run-time. There are other overheads associated with this method: the frames can become bigger and hence slow down the SAT solver, and the check in line 11, which has a complexity linear in the size of the frame, can also be negatively affected. All these factors show that without careful tuning, this technique is not likely to succeed.

We implemented our suggested algorithms on top of SimpleCAR [16,23], which is an implementation of the CAR algorithm. The following table shows the effect of applying this technique on the total running time and the total number of SAT calls (in line 18). The timeout here and in the rest of the article was set to one hour.

---

[3] Results of these benchmarks are either known to be unsafe or remain unknown. Those that are already proven to be safe in the competition are excluded, as we will explain in Sect. 5.

[4] The focus on these relatively old benchmarks stems from the fact that since HWMCC 2019, the official format shifted to BTOR, which operates at the word level. Although some bit-level benchmarks were provided, they relied on a newer version of the Aiger format (> 1.9). However, not all the model checkers that we compared against support this format, and some emit wrong results. Consequently, we excluded those benchmarks and focused on HWMCC 2015 and HWMCC 2017 instead. We note that the most recent bit-level model checking article that was published to the best of our knowledge [26] also uses the 2015 and 2017 benchmarks.

Table 2. The effect of adding more UCs using different ordering of assumptions. The subscript in mUC_r indicates that the assumptions orderings for generating the additional cores were selected randomly.

	mUC (1)	mUC (2)	$mUC_r(2)$	$mUC_r(3)$	$mUC_r(4)$	$mUC_r(5)$
Solved	146	**150**	146	142	142	139
Avg. # of (first-only) SAT calls	137k	103k	146k	120k	107k	97k
Avg. total time of 1st SAT calls	1171.2 s	944.0 s	1138.7 s	1014.0 s	879.7 s	827.9 s
Avg. total time of subsequent SAT calls	0.0 s	367.3 s	741.4 s	916.7 s	1041.2 s	1145.6 s
Avg. UC redundant rate	0	4.53%	7.09%	8.97%	10.70%	11.15%

As we can see from Table 2, the number of cases solved by mUC (2) increased, which is a success. Furthermore, the number of (first) SAT calls and their accumulated run-time indeed drops, which implies that we are able to narrow down the over-approximation by adding more UCs, as expected. Note that the total SAT run-time (944+367.3 s) of mUC (2) is larger than that of the baseline mUC (1) – 1171.2 s On the other hand, the table shows that adding cores based on random assumptions ordering does not help. We will therefore continue from here on with mUC (2) based on the reverse order.

The results in the table may be misleading due to the time-out cases. For example, if a run with a single UC terminates on time and manages to make many fast SAT calls, while a run with two UCs times out and, due to overhead, completes fewer SAT calls before timing out, the table would count more calls under the single UC column. This could create the false impression that the single UC approach is less efficient than the two UCs approach. To address this, we also calculated the number of initial SAT calls after excluding 173 cases where at least one engine timed out:

	mUC (1)	mUC (2)
Avg. # of (first-only) SAT calls	2960	1747

We can see, then, that for this set of instances, the number of (first) SAT calls also drops. Notably this filtered view can also distort the full story, exactly because it ignores the cases in which one terminates and the other does not.

When comparing the overall runtime, the individual results of each benchmark instance in Fig. 1 show that most of the plots are above the diagonal, indicating a positive impact of adding the second UC.

4.2   Re-tuning the Solver for Multiple UCs

Selective Application. To try to improve the method further, we experimented with various heuristics that control when to apply it. Specifically, in which frames to apply it: the high or low $x\%$ of the frames. For example, if we

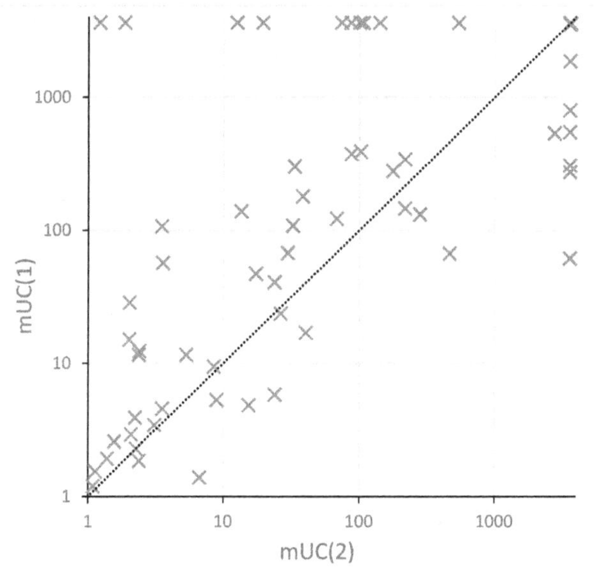

Fig. 1. Per-instance comparison of total time where X-axis and Y-axis represents mUC(2) and mUC(1) respectively.

choose 30% from the low-index frames, it means that if the current number of O frames is n, then for frame i, $i \leq 0.3n$, we add multiple cores. Table 3 shows our results with different such values.

Table 3. Different options to add multiple UCs

	mUC (1)	mUC (2)	Low-*				High-*			
			50%	33%	25%	20%	50%	33%	25%	20%
Solved	146	150	151	**152**	150	149	149	150	147	144

Hence, by adding a UC only for the 33% low-index frames we are able to solve two more cases.

Detecting Blocked States Efficiently. Next, through profiling we learned that a large part of the run-time is spent in line 11 of Algorithm 1, which tests whether the state s is blocked by the UCs in the frame O_{l+1}. Recall that this step is an optimization that saves SAT calls in line 18. The time spent in this line is proportional (i.e., linear) to the size of the frame, and hence adding more UCs to it potentially increases the overhead of this test. We measured the portion of running time dedicated to this process in the time-out cases. On average, it was 11% (407 out of 3600 sec). In 14% of the cases it was more than 30% of the running time. Algorithm 2 shows the implementation of 'blockedIn'. It iterates the clauses in the corresponding frame and checks if one of them blocks the state, which amounts to a subsumption test.

Algorithm 2: The Basic Implementation of 'blockedIn'

Input: A state s, an O frame O_l
Output: 'BLOCKED' or 'NOT BLOCKED'

1 **for** *each $cl \in O_l$* **do**
2   **if** $cl \rightarrow \neg s$ **then**
3   return 'BLOCKED'

4 return 'NOT BLOCKED'

Going over all clauses in sequence might not be the best strategy. We experimented with using BCP instead, which amounts to calling a SAT solver over the formula O_l and the assumptions s. Since s spans the entire set of variables, this SAT call is determined without real decisions (i.e., only decisions of the assumptions themselves) and is hence linear. Furthermore, we maintain a separate solver object for each frame, and hence we activate this step incrementally, i.e., we do not need to read the frame into the SAT solver each time.

While Algorithm 2 can be thought of as clause-driven DFS (checking each clause to completion each time), the BCP method can be thought of as literal-driven BFS. The 2-watch literal scheme [19] in modern SAT solvers makes it highly efficient, as most clauses are not even visited, so we expect this route to be faster, at least when the frames are large. Our experiments confirmed this hypothesis, which led us to a *hybrid* approach based on a threshold: we activate the BCP method only when the number of clauses in the frame is larger than 10k.

With this threshold, our results improved by three solved cases, to 155—see Table 4. Two things to note in this table: first, the hybrid approach of invoking Algorithm 2 for frames with 10k clauses or less, and BCP otherwise, improves the results regardless of the number of added UCs; Second, the time per call decreased by a factor of 2 (from 0.53 to 0.26) with the latest configuration, owing to this method.

Restarts. Finally, we tested whether restarting periodically with different mUC strategies can help. Each such strategy likely leads to a different search. At each restart, we preserve some data from the previous run. It is important to note that we cannot maintain arbitrary portions of the frames, because it may break the invariants listed in Definition 1. For example, if we remove a clause from O_{k_o}, then we can no longer guarantee that the induction condition is maintained, namely that $O_{k+1} \supseteq R^{-1}(O_k)$. If we arbitrarily remove a state from a U frame, say U_k, we cannot guarantee that the invariant $U_{k+1} \subseteq R(U_k)$ is maintained, although this invariant is not necessary for the proof of correctness [17].

The only correct and computationally easy way that we found to remove data from the O sequence is to decide on an index k_o such that $O_0, \ldots, O_{k_o}$ are

Table 4. Results with different 'blockedIn' approaches. The 'Basic' rows refer to Algorithm 2.

UC method	blockedIn method	Avg time per call (ms)	Solved
mUC (1)	Basic	0.56	146
	BCP	0.81	145
	Hybrid (10k)	0.4	148
mUC (2)	Basic	0.96	150
	BCP	0.85	150
	Hybrid (10k)	0.52	**153**
mUC (2) (Low 33%)	Basic	0.53	152
	BCP	0.58	153
	Hybrid (10k)	0.26	**155**

maintained whereas O frames with a larger index are removed. One exception that can be made to this rule is with the border frame, O_{k_o}, that can also be *partially* removed without violating the invariant, because after resetting O_{k_o+1} it is equal to *true* and trivially maintains the invariant. Similarly, for the U sequence, we can decide on an index k_u, maintain $U_0, \ldots, U_{k_u}$ and remove frames with an index larger than k_u.

We experimented with changing five dimensions:

1. The values of k_o, k_u, namely the border frames that define which data to maintain between restarts,
2. the portion of the border frames to maintain,
3. the clauses to be maintained in the border frames,
4. the selective application strategy (see the beginning of this subsection)
5. the time to restart.

The best configuration that we found is with $k_o = 1, k_u = 0$. As for the second question, we settled on maintaining at the nth restart $\frac{n}{n+1}$ of O_{k_o}. Hence, we begin with half of O_{k_o}, then two-thirds, and so on, namely we increase it from one restart to the next. Note that O_{k_o} likely changes each time. As for the third question, we sort O_{k_o}'s clauses and take the shortest ones. As for the fourth question, we begin with full activation and then begin to only activate at the low indices, where the activation threshold is updated at restart n, for $n > 1$, according to

$$threshold[n] = \frac{threshold[n-1]}{threshold[n-1]+1} \tag{1}$$

Hence the activation ratio progresses as follows: $1, 1/2, 1/3, 1/4, \ldots$. Finally, as for the fifth question, we experimented with different time constants.

Table 5 summarizes some of our experiments with these dimensions. We can now solve more than 160 cases with many different configurations, peaking at 164 cases.

Table 5. The number of solved cases by mUC (2) with different restart strategies and restarting periods.

Restart Strategy	$t=180\,$s	$t=300\,$s	$t=450\,$s	$t=600\,$s	$t=900\,$s	$t=1200\,$s
$k_o = 0, k_u = 0$	155	157	160	156	155	160
$k_o = 1, k_u = 0$	154	157	157	159	161	158
$k_o = 1, k_u = 0$, forget parts of O_1	156	**164**	161	157	158	160

5   A Comparison to Other Model Checkers

Finally, we compared mUC to the leading model checkers in the public domain: ABC-BMC [1], ABC-PDR [1], NUXMV-IGOOD [21,26], AVY [2,24], NAVY [20,25], BAC [27] and our baseline tool SIMPLECAR [17,23]. Details about the configuration of these tools in our experiments appear in the appendix. The benchmark set is the same as described in Sect. 4.1.

We ran the experiments on a cluster of Linux servers, each equipped with an Intel Xeon Gold 6132 14-core processor at 2.6 GHz and 96 GB RAM. The version of the operating system is Red Hat 4.8.5-16. For each running instance, the number of CPU cores was limited to 1, the memory was limited to 8 GB, and the time to 1 h.

Table 6 shows a summary of this comparison. The last column refers to mUC in its best configuration as described in the previous section. Figure 2 and Table 7 give more details about the running time.

Table 6. The number of unsafe instances solved with a 1-hour timeout, by different model checkers.

	ABC-BMC	ABC-PDR	NUXMV-IGOOD	AVY	NAVY	BAC	SIMPLECAR	mUC
Solved	159	109	139	128	138	160	146	**164**

Table 7. Comparing mUC to other model checkers in terms of the percentage of benchmarks that they each solve at least 5 s faster than the other.

	ABC-BMC	ABC-PDR	NUXMV-IGOOD	AVY	NAVY	BAC	SIMPLECAR
mUC is faster	34.0%	**63.4%**	**71.3%**	**58.2%**	**48.2%**	**61.2%**	**35.5%**
mUC is slower	**43.3%**	9.2%	13.3%	13.5%	20.6%	12.7%	14.0%

During our experiments with the various configurations of mUC, we were able to solve 7 benchmarks that were in an *unknown* status so far, that no *current version* of the tools in Table 6 can solve within the time limit, thus this can be

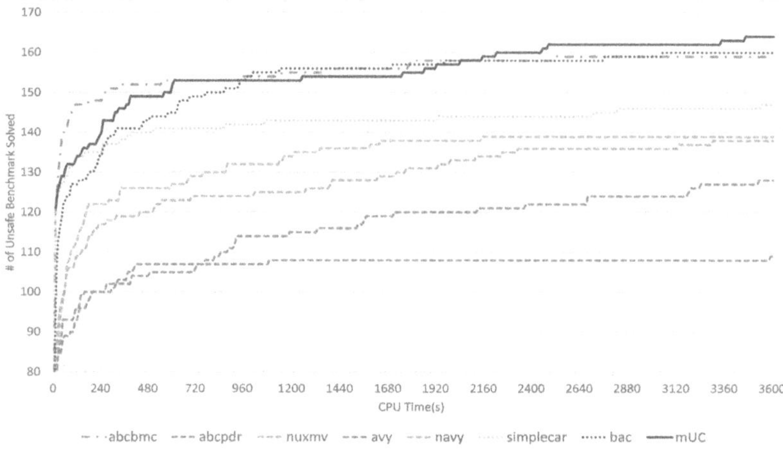

Fig. 2. Comparison of run-time performance among different model checkers (best seen in color). (Color figure online)

seen as a contribution to the state-of-the-art. The depths of the counterexamples that we found, compared to the maximal depths that were proven safe by the virtual best solver of the other tools, appear in Table 8. We note that in some of the cases, the configuration at the left column is only one of several configurations that solved the benchmark.

Table 8. Benchmarks that only mUC can solve, albeit under different configurations (in some cases by multiple mUC configurations). The third column represents the maximum depth that other tools reached together (vbs), i.e., were able to prove that there is no counterexample up to that depth. The right column is the length of the counterexample that mUC found.

mUC (2) configuration	Benchmark	Bound	Counterexample length by mUC
low-20%	oski15a01b76s	14	143
low-25%	oski15a01b62s	16	95
low-33%	oski15a01b70s	18	125
low-33%	6s329rb20	47	1417
low-50%	intel012	100	1962
high-33%	6s329rb19	45	1295
full	oski15a01b36s	14	262

For completeness, we also report the results for safe cases in Table 9, despite the fact that CAR in general is not competitive for safety checking. The results show that our technique also improves the results for those cases—in particular from 166 to 177 solved instances.

Table 9. The number of safe instances solved (out of the 749 instances of HWMCC15-17) with a 1-hour timeout, by different model checkers.

	ABC-BMC	ABC-PDR	NUXMV-IGOOD	AVY	NAVY	BAC	SIMPLECAR	mUC
Solved	-	294	304	296	310	-	166	177

To summarize, our experiments show that our implementation of these techniques, on top of SIMPLECAR, not only improves its performance, but also solves more (unsafe) cases than any other model-checker in the public domain, and solves instances that no other model checker can.

6   Conclusion

SAT-based model checkers like CAR and PDR rely heavily on unsatisfiable cores to generalize from failed attempts to progress from a given state. While previous work [10] focused on improving the core by controlling the literal order, here we suggested to add several cores (although in practice we found that just adding one more core, and even that only selectively, is best). Our initial core was already the best one (as a single core!) according to [10], hence our technique improves it. We discussed the many dimensions of this technique and presented our (limited) experience with tuning them, although there is plenty left for future research. For example, what are good literal orders for achieving high-quality cores in the context of adding multiple cores (we simply took the order suggested in [10] and its reversal for the 2nd core)? What is the best set of U frames and partial frames to leave when restarting? Can this method also improve PDR?

We showed that despite the fact that additional cores can be computed in linear time, and it indeed saves normal, worst-case exponential SAT runs, owing to the fact that the latter process is fed with relatively easy formulas, it is not at all trivial that the saving will be larger than the overhead of computing more cores and the overhead associated with having larger frames. We suggested a method for improving the linear-time test called 'blockedIn' (line 11 of Algorithm 32) which initially was made slower by the increased frames, and also showed that restarting the solver with different strategies improves the run time.

At the bottom line, we were able to improve CAR from 146 to 164 solved cases. The next in line is BAC, which is also based on CAR, and after that the bounded model checker ABC-BMC, which can only solve 160 and 159 cases, respectively. Furthermore, we improved the state-of-the-art in the sense that we solved 7 cases that no other solver can. Such an improvement is significant in light of the many years of research on these tools.

All the artifacts described in this article are available at [9].

A   The Configurations of the Competing Model Checkers

The table below gives more details about the configuration of each of the tools. Unless otherwise stated, these flags represent the default. In the case of AVY and

NAVY the flags were copied from the corresponding scripts in their respective repository.

Tool	Configuration Flags
ABC-BMC	-c "bmc2"
ABC-PDR	-c "pdr"
SIMPLECAR	-b -e
BAC	-b -e -1500
AVY	–reset-cover=1 -a –kstep=1 –shallow-push=1 –tr0=1 –min-suffix=1 –glucose –glucose-inc-mode=0 –min-core=1 –glucose_itp=1 –stick-error=1 –sat-simp=1
NAVY	–reset-cover=1 –opt-bmc –kstep=1 –shallow-push=1 –min-suffix=1 –glucose –glucose-inc-mode=0 –sat-simp=1 –glucose_itp=1
NUXMV-IGOOD	-a ic3 -s cadical -W -m 1 -u 4 -I 1 -D 0 -g 1 -X 0 -c 0 -p 1 -d 2 -G 1 -P 1 -A 100 -O 3
MUC	–vb –inter 1 –rotate –raw –imp 5 –rem 1 –convMode 1 –convParam 0 –restart 300

References

1. ABC: System for Sequential Logic Synthesis and Formal Verification. https://github.com/berkeley-abc/abc
2. AVY. https://arieg.bitbucket.io/avy/
3. Bernardini, A., Ecker, W., Schlichtmann, U.: Where formal verification can help in functional safety analysis. In: IEEE/ACM International Conference on Computer-Aided Design (ICCAD), pp. 1–8 (2016). https://doi.org/10.1145/2966986.2980087
4. Biere, A., Cimatti, A., Clarke, E., Fujita, M., Zhu, Y.: Symbolic model checking using SAT procedures instead of BDDs. In: Proceedings of Design Automation Conference (DAC), pp. 317–320 (1999). https://doi.org/10.1109/DAC.1999.781333
5. Biere, A.: AIGER Format. http://fmv.jku.at/aiger/FORMAT
6. Biere, A., Cimatti, A., Clarke, E., Zhu, Y.: Symbolic model checking without BDDs. In: Cleaveland, W.R. (ed.) Tools and Algorithms for the Construction and Analysis of Systems (TACAS), pp. 193–207. Springer, Berlin, Heidelberg (1999)
7. Bradley, A.R.: SAT-based model checking without unrolling. In: Jhala, R., Schmidt, D. (eds.) Verification, Model Checking, and Abstract Interpretation, pp. 70–87. Springer, Berlin, Heidelberg (2011)
8. Burch, J., Clarke, E., McMillan, K., Dill, D., Hwang, L.: Symbolic model checking: 10^{20} states and beyond. Inf. Comput. **98**(2), 142–170 (1992). https://doi.org/10.1016/0890-5401(92)90017-A
9. Dong, Y.: mUC-artifact (2025). https://doi.org/10.5281/zenodo.15227232

10. Dureja, R., Li, J., Pu, G., Vardi, M.Y., Rozier, K.Y.: Intersection and rotation of assumption literals boosts bug-finding. In: Chakraborty, S., Navas, J.A. (eds.) Verified Software. Theories, Tools, and Experiments, pp. 180–192. Springer International Publishing, Cham (2020)
11. Een, N., Mishchenko, A., Brayton, R.: Efficient implementation of property directed reachability. In: Proceedings of the International Conference on Formal Methods in Computer-Aided Design, pp. 125–134. FMCAD '11, FMCAD Inc, Austin, Texas (2011)
12. Eén, N., Sörensson, N.: An extensible SAT-solver. In: Giunchiglia, E., Tacchella, A. (eds.) Theory and Applications of Satisfiability Testing, pp. 502–518. Springer, Berlin, Heidelberg (2004)
13. HWMCC 2024. https://hwmcc.github.io/2024/
14. IC3Ref. https://github.com/arbrad/IC3ref
15. Jhala, R., Majumdar, R.: Software model checking. ACM Comput. Surv. (CSUR) **41**(4), 1–54 (2009). https://doi.org/10.1145/1592434.1592438
16. Li, J., Dureja, R., Pu, G., Rozier, K.Y., Vardi, M.Y.: SimpleCAR: an efficient bug-finding tool based on approximate reachability. In: Chockler, H., Weissenbacher, G. (eds.) Computer Aided Verification, pp. 37–44. Springer International Publishing, Cham (2018). https://doi.org/10.1007/978-3-319-96142-2_5
17. Li, J., Zhu, S., Zhang, Y., Pu, G., Vardi, M.Y.: Safety model checking with complementary approximations. In: 2017 IEEE/ACM International Conference on Computer-Aided Design (ICCAD), pp. 95–100 (2017). https://doi.org/10.1109/ICCAD.2017.8203765
18. McMillan, K.L.: Interpolation and SAT-based model checking. In: Hunt, W.A., Somenzi, F. (eds.) Computer Aided Verification, pp. 1–13. Springer, Berlin, Heidelberg (2003). https://doi.org/10.1007/978-3-540-45069-6_1
19. Moskewicz, M., Madigan, C., Zhao, Y., Zhang, L., Malik, S.: Chaff: engineering an efficient SAT solver. In: Proceedings of the Design Automation Conference (DAC'01) (2001)
20. NAVY. https://bitbucket.org/arieg/extavy/downloads/fib_cav15.tar.gz
21. nuXmv with i-good Lemmas. https://github.com/youyusama/i-Good_Lemmas_MC
22. Seufert, T., Scholl, C., Chandrasekharan, A., Reimer, S., Welp, T.: Making PROGRESS in property directed reachability. In: Finkbeiner, B., Wies, T. (eds.) Verification, Model Checking, and Abstract Interpretation, pp. 355–377. Springer International Publishing, Cham (2022). https://doi.org/10.1007/978-3-030-94583-1_18
23. SimpleCAR. https://github.com/lijwen2748/simplecar/releases/tag/v0.1 (2018)
24. Vizel, Y., Gurfinkel, A.: Interpolating property directed reachability. In: International Conference on Computer Aided Verification, pp. 260–276. Springer (2014)
25. Vizel, Y., Gurfinkel, A., Malik, S.: Fast interpolating BMC. In: Computer Aided Verification: 27th International Conference, CAV 2015, San Francisco, CA, USA, July 18-24, 2015, Proceedings, Part I 27, pp. 641–657. Springer (2015)
26. Xia, Y., Becchi, A., Cimatti, A., Griggio, A., Li, J., Pu, G.: Searching for i-good lemmas to accelerate safety model checking. In: International Conference on Computer Aided Verification, pp. 288–308. Springer (2023)
27. Zhang, X., Xiao, S., Li, J., Pu, G., Strichman, O.: Combining BMC and complementary approximate reachability to accelerate bug-finding. In: Proceedings of the 41st IEEE/ACM International Conference on Computer-Aided Design. ICCAD '22, Association for Computing Machinery, New York, NY, USA (2022). https://doi.org/10.1145/3508352.3549393

Efficient Model Checking
for the Alternating-Time μ-Calculus
via Effectivity Frames

Daniel Hausmann[1], Merlin Humml[2], Simon Prucker[2(✉)], and Lutz Schröder[2]

[1] University of Liverpool, Liverpool, UK
[2] Friedrich-Alexander-Universität Erlangen-Nürnberg, Erlangen, Germany
`simon.prucker@fau.de`

Abstract. The semantics of alternating-time temporal logic (ATL) and the more expressive alternating-time μ-calculus (AMC) is standardly given in terms of concurrent game frames (CGF). The information required to interpret AMC formulas is equivalently represented in terms of effectivity frames in the sense of Pauly; in many cases, this representation is more compact than the corresponding CGF, and in principle allows for faster evaluation of coalitional modalities. In the present work, we investigate whether implementing a model checker based on effectivity frames leads to better performance in practice. We implement the translation from concurrent game frames to effectivity frames and analyse performance gains in model checking based on corresponding instantiations of a generic model checker for coalgebraic μ-calculi, using dedicated benchmark series as well as random systems and formulas. In the process, we also compare performance to the state-of-the-art ATL model checker MCMAS. Our results indicate that on large systems, the overhead involved in converting a CGF to an effectivity frame is often outweighed by the benefits in subsequent model checking.

Keywords: Model checking · multi-agent systems · alternating-time temporal logic · concurrent game frames

1 Introduction

Alternating-time temporal logic (ATL) and its extension with full fixpoints, the alternating-time μ-calculus (AMC) play an established role as a core formalism for the specification of multi-agent systems [2]. They allow expressing the ability

D. Hausmann—Supported by the ERC Consolidator grant D-SynMA (No. 772459) and by the EPSRC through grant EP/Z003121/1.

M. Humml—Funded by the German Federal Chamber of Notaries, project DIREGA.

S. Prucker—Funded by the Deutsche Forschungsgemeinschaft (DFG, German Research Foundation) – project number 517924115.

L. Schröde—Funded by the Deutsche Forschungsgemeinschaft (DFG, German Research Foundation) – project number 419850228.

G. Ernst and K. Y. Rozier (Eds.): SPIN 2025, LNCS 15945, pp. 106–124, 2026.
https://doi.org/10.1007/978-3-032-06847-7_6

of groups of agents (*coalitions*) to achieve short-term and long-term goals by coordinated action. The semantics of ATL and the AMC is standardly defined in terms of *concurrent game frames* (CGFs), which at each state assign to each agent an explicit set of available moves, along with an outcome function determining which successor state is reached once all agents have picked a move. It has been shown that for purposes of evaluating AMC formulas, one may equivalently convert a CGF into an *effectivity frame*, which at each state records, for each coalition C, a set of sets of states, understood as consisting of those state properties that C may (alternatively) enforce by joint action in the next state [20].

Like in the case of standard temporal logics, one of the key verification tasks in this context is *model checking*, i.e. to determine whether a given state in a multi-agent system, say a CGF, satisfies a given formula of ATL or the AMC. The most natural way to present the input system for purposes of ATL or AMC model checking is, presumably, indeed to specify a CGF, which makes the actions available to the agents and their effects explicit. On the other hand, it is immediate from inspection of the respective semantics that evaluation of alternating-time modalities, which over CGFs amounts to the evaluation of nested quantifiers over actions ("coalition C has a joint action such that no matter what the other agents do, the target formula holds") and hence takes exponential time in the number of agents, is computationally more efficient on effectivity frames, where it reduces to a simple look-up operation.

In the present work, we explore whether this observation can be translated into actual efficiency gains in model checking. That is, we trade the computational cost of evaluating coalitional modalities on CGFs for an additional preprocessing step in which we translate a CGF into an effectivity frame. The computations involved in the translation are, prima facie, similar to the ones involved in evaluating coalitional modalities on CGFs, but they are performed only once, while especially in the AMC, modalities may be evaluated repeatedly at the same state. We provide an implementation of AMC model checking both on CGFs and on effectivity frames, building on a generic implementation of model checking algorithms for coalgebraic μ-calculi [11] within the *Coalgebraic Ontology Logic Solver (COOL)* [8–10]. Thereby, we incidentally provide, as far as we are aware, the first model checker that covers the entire AMC (which contains ATL* as a fragment [2, Theorem 6.1]), rather than only ATL; furthermore, our tool is agnostic as to whether the input multi-agent system is provided in the form of a CGF or as an effectivity frame and hence enables working directly with effectivity frames. We run extensive experiments both on random models and formulas and on dedicated benchmark series; besides comparing between CGF-based and effectivity-frame-based model checking, we also compare the performance of our implementation with the state-of-the-art ATL model checker MCMAS [17]. As expected, results vary strongly with the exact nature of benchmark series, in particular in the comparison with MCMAS; notably, MCMAS implements symbolic model checking while COOL offers only explicit-state model checking, and correspondingly

MCMAS performs better on problems that can be succinctly represented symbolically while COOL plays out advantages when this is not the case.

Related Work. As indicated above, the notion of effectivity frame semantics was introduced by Pauly [20]. Goranko and Jamroga [6] provide a comparison of ATL semantics over CGFs, effectivity frames, and *alternating transition systems* (ATSs), in which the moves of the agents are subsets of the state space. The translation from CGFs to ATSs involves a blowup of the state space, so we refrain from including ATS semantics in the present comparison. As far as we are aware, the first implemented model checker for ATL was MOCHA [1], which however seems to be no longer available. Kański et al. [13] compare MCMAS to an ATL model checker UMC4ATL based on McMillan's method of *unbounded model checking* [18]. The results indicate better performance of MCMAS, on which we therefore focus in our comparative evaluation. The complexity of model checking ATL and related logics has been investigated extensively already at the time of introduction [2]; we additionally mention work on symbolic model checking for ATL [12] and on the complexity of explicit-state model checking for ATL$^+$ [7]. The AMC (hence also ATL*) is subsumed by the more expressive strategy logic (SL) [5], which allows the specification of strategies that fail to be ω-regular, but has nonelementary model checking [19]; restricted fragments of SL with more favourable model checking complexities have been considered [3, 4] (and implemented in MCMAS-SLK [4]).

2   The Alternating-Time μ-Calculus

We briefly recall the syntax and semantics of the *alternating-time μ-calculus*, a fixpoint logic that allows the specification of ω-regular joint strategies of coalitions in multi-agent systems [2]. Its key feature are *coalitional modalities* $[C]$ 'coalition C of agents is able to enforce'; temporal idioms over this base are expressed using least and greatest fixpoint operators. The alternating-time μ-calculus contains *alternating-time temporal logic (ATL)* as a fragment, embedded in essentially the same way as computational tree logic embeds into the standard μ-calculus.

Syntax. Formulas of the *alternating-time temporal μ-calculus (AMC)* are given by the following grammar, depending on (countable) sets Ag, At and V of *agents*, *propositional atoms* and *fixpoint variables*, respectively.

$$\varphi, \psi ::= \top \mid \bot \mid p \mid \neg p \mid \varphi \wedge \psi \mid \varphi \vee \psi \mid [C]\varphi \mid \langle C \rangle \varphi \mid X \mid \eta X. \varphi$$

where $p \in$ At, $X \in$ V, $\eta \in \{\mu, \nu\}$, and C ranges over *coalitions*, i.e. subsets of Ag. *Coalitional modalities* $[C]\varphi$ (for which we employ notation as in *coalition logic* [20]) intuitively express that the agents in coalition C have a joint strategy to enforce φ in the next step of the game; dually, $\langle C \rangle \varphi$ expresses that coalition C cannot prevent that φ is satisfied in the next step. The fixpoint operators give rise to standard notions of *bound* and *free* (occurrences of) fixpoint variables;

a formula is *closed* if it does not contain free fixpoint variables, and *clean* if every fixpoint variable in it is bound by at most one fixpoint operator. Given a closed formula φ, we let $|\varphi|$ denote its syntactic size. The algorithms and implementations in this paper work on the *closure* $\mathsf{cl}(\varphi)$ of φ. The closure is a succinct graph representation of the respective formula, intuitively obtained from its syntax tree by identifying occurrences of fixpoint variables with their binding fixpoint operators; we have $|\mathsf{cl}(\varphi)| \leq |\varphi|$. Finally, we define the *alternation-depth* $\mathsf{ad}(\varphi)$ of fixpoint formulas $\varphi = \eta X. \psi$ in the usual way, capturing the number of alternations between least and greatest fixpoints; for a detailed account of these syntactic notions, see [15].

As mentioned in Sect. 1, the AMC has various equivalent semantics; we recall the semantics based on *concurrent game frames* [2] and that based on *effectivity frames* [20]. Let W be a non-empty finite set of states. A model $\mathcal{M} = (\mathcal{F}, \rho)$ consists of a *valuation* $\rho : \mathsf{At} \to \mathcal{P}(W)$ that assigns to each propositional atom $a \in \mathsf{At}$ the set $\rho(a)$ of states that satisfy a; $\mathcal{F}$ is either a concurrent game frame (over W) or an effectivity frame (over W), introduced next.

Concurrent Game Frame Semantics: For a coalition C, we write $\Pi_C = \mathbb{N}^{|C|}$ to denote the set of *joint (one-step) strategies* (or *joint moves*) of C; we denote the set Π_{Ag} of *complete (one-step) strategies* (or *grand moves*) by just Π. Given a joint strategy $s_C \in \Pi_C$ of C, any joint strategy $s_{\overline{C}} \in \Pi_{\overline{C}}$ of the counter coalition $\overline{C} = \mathsf{Ag} \setminus C$ induces a complete strategy $(s_C, s_{\overline{C}}) \in \Pi$. In this notation, a *concurrent game frame (CGF)*

$$\mathcal{F} = (W, m : W \times \mathsf{Ag} \to \mathbb{N}, f : W \times \Pi \rightharpoonup W)$$

consists of W together with functions

- $m : W \times \mathsf{Ag} \to \mathbb{N}$, assigning to each state $w \in W$ and each agent $a \in \mathsf{Ag}$ the number $m(w, a) > 0$ of moves available to agent a at the state w; and
- $f : W \times \Pi \rightharpoonup W$, assigning to each state $w \in W$ and each complete strategy $s = (s_1, \ldots, s_{|\mathsf{Ag}|}) \in \Pi$ that is *admissible* at w, i.e. $s_i < m(w, i)$ for all $i \in \mathsf{Ag}$, an *outcome* state $f(w, s) \in W$.

Satisfaction $w, \sigma \models \varphi$ of an AMC formula φ in a state w of a CGF $\mathcal{F} = (W, m, f)$ is defined inductively, depending on a *valuation* $\sigma : \mathsf{V} \to \mathcal{P}(W)$ of the fixpoint variables; we denote the *extension* of φ under σ by $\llbracket \varphi \rrbracket_\sigma = \{w \in W \mid w, \sigma \models \varphi\}$. For a coalition C and a state $w \in W$, we let Π_C^w denote the set of joint strategies of C that are admissible at w. The semantic clauses for propositional atoms and connectives are as usual (the former is given by the valuation ρ of atoms); modalities are interpreted as

$$\llbracket [C]\varphi \rrbracket_\sigma = \{w \in W \mid \exists s_C \in \Pi_C^w. \forall s_{\overline{C}} \in \Pi_{\overline{C}}^w. f(w, (s_C, s_{\overline{C}})) \in \llbracket \varphi \rrbracket_\sigma\}$$

$$\llbracket \langle C \rangle \varphi \rrbracket_\sigma = \{w \in W \mid \forall s_C \in \Pi_C^w. \exists s_{\overline{C}} \in \Pi_{\overline{C}}^w. f(w, (s_C, s_{\overline{C}})) \in \llbracket \varphi \rrbracket_\sigma\}$$

(where we write $(s_C, s_{\overline{C}})$ for the complete strategy extending s_C and $s_{\overline{C}}$) so that, e.g., $w, \sigma \models [C]\varphi$ if and only if coalition C has a joint strategy s_C at w such that for all strategies that complete s_C and are admissible at w, the outcome satisfies φ under the valuation σ of fixpoint variables; and the semantics of fixpoint operators is given (exploiting the Knaster-Tarski fixpoint theorem) by

$$\llbracket \mu X.\,\varphi \rrbracket_\sigma = \bigcap \{Z \subseteq W \mid \llbracket \varphi \rrbracket_\sigma^X(Z) \subseteq Z\} \quad \llbracket \nu X.\,\varphi \rrbracket_\sigma = \bigcup \{Z \subseteq W \mid Z \subseteq \llbracket \varphi \rrbracket_\sigma^X(Z)\},$$

where $\llbracket \varphi \rrbracket_\sigma^X$ is the function defined by $\llbracket \varphi \rrbracket_\sigma^X(A) = \llbracket \varphi \rrbracket_{\sigma[X \mapsto A]}$ for $A \subseteq W$, where $\sigma[X \mapsto A](X) = A$ and $\sigma[X \mapsto A](Y) = \sigma(Y)$ for $Y \neq X$. For a *closed* formula φ (in which all fixpoint variables X occur in the scope of some binding fixpoint operator μX or νX), $\llbracket \varphi \rrbracket_\sigma$ does not depend on σ, so in this case we just write $\llbracket \varphi \rrbracket$ (or $\llbracket \varphi \rrbracket_{\mathcal{F}}$ to make the semantic domain explicit) for $\llbracket \varphi \rrbracket_\sigma$.

Effectivity Frame Semantics: An *effectivity frame (EF)*

$$\mathcal{F} = (W, e\colon W \times \mathcal{P}(\mathsf{Ag}) \to \mathcal{P}(\mathcal{P}(W)))$$

consists of W together with an effectivity function e that assigns, to each state $w \in W$ and each coalition $C \subseteq \mathsf{Ag}$, a set $e(w, C)$ of sets of states. Intuitively, we will have $U \in e(w, C)$ whenever coalition C has a joint strategy s_C at w that guarantees that the next state in the game is contained in U. To support this intuitive meaning, effectivity frames need to be restricted to be *playable*, ensuring for finite W that they are induced by a CGF [6,16,20]. When constructing effectivity frames from CGFs, playability is automatic.

Satisfaction of AMC formulas over an effectivity frame $\mathcal{F} = (W, e)$ is defined inductively by the same clauses for propositional and fixpoint operators as for CGFs, but modalities are interpreted as

$$\llbracket [C]\varphi \rrbracket_\sigma = \{w \in W \mid \exists U \in e(w, C).\, U \subseteq \llbracket \varphi \rrbracket_\sigma\}$$
$$\llbracket \langle C \rangle \varphi \rrbracket_\sigma = \{w \in W \mid \forall U \in e(w, C).\, U \cap \llbracket \varphi \rrbracket_\sigma \neq \emptyset\}$$

so that with this semantics, $w, \sigma \models [C]\varphi$ if and only if coalition C can enforce some set U at state w such that all states contained in U satisfy φ (under the valuation σ of fixpoint variables). The difference between the two semantics is displayed in Fig. 1, which illustrates the encoding of a single game step in both representations.

In the CGF encoding on the left hand side, we have, e.g., $w_1 \models [\{1,3\}]q$ since, for the joint strategy that agents 1 and 3 both pick their second move, the outcome is always w_3, no matter which move agent 2 picks. This is reflected in the EF encoding by $\{1, 3\}$ being able to force the singleton set $\{w_3\}$.

Concurrent game frames can be transformed into equivalent effectivity frames [20]; explicitly, the conversion works as follows.

Definition 1 (Induced effectivity frame). A CGF $\mathcal{F} = (W, m, f)$ *induces* the effectivity frame $\mathcal{F}' = (W, e)$ defined by

$$e(w, C) = \{\{f(w, s_C, s_{\overline{C}}) \mid s_{\overline{C}} \in \Pi_{\overline{C}}^w\} \mid s_C \in \Pi_C^w\}$$

for $w \in W$ and $C \subseteq \mathsf{Ag}$.

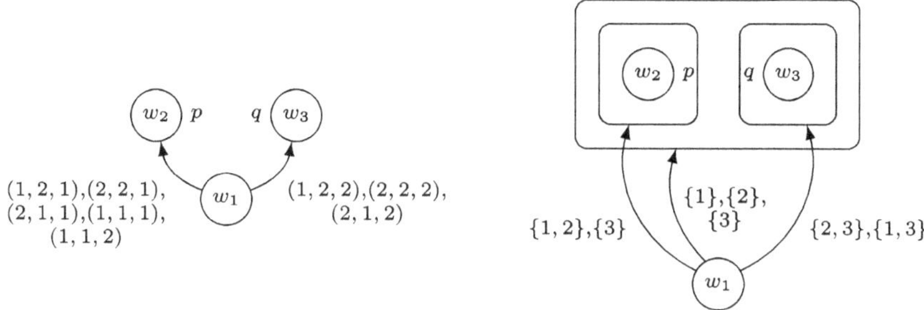

Fig. 1. Concurrent game frame semantics vs. effectivity frame semantics.

Lemma 1. *Let $\mathcal{F}$ be a concurrent game frame and let $\mathcal{F}'$ be the effectivity frame induced by $\mathcal{F}$, and let σ be a valuation. Then for all formulas φ,*

$$[\![\varphi]\!]_{\mathcal{F},\sigma} = [\![\varphi]\!]_{\mathcal{F}',\sigma}.$$

Proof. The claim follows immediately from the equi-satisfaction of modalities in $\mathcal{F}$ and $\mathcal{F}'$. For readability, we show this for closed modal formulas $[C]\psi$. Assuming that $[\![\psi]\!]_{\mathcal{F}} = [\![\psi]\!]_{\mathcal{F}'}$, we have

$$
\begin{aligned}
[\![[C]\psi]\!]_{\mathcal{F}} &= \{w \in W \mid \exists s_C \in \Pi_C^w. \forall s_{\overline{C}} \in \Pi_{\overline{C}}^w. f(w,(s_C,s_{\overline{C}})) \in [\![\psi]\!]_{\mathcal{F}}\} \\
&= \{w \in W \mid \exists s_C \in \Pi_C^w. \forall u \in \{f(w,s_C,s_{\overline{C}}) \mid s_{\overline{C}} \in \Pi_{\overline{C}}^w\}. u \in [\![\psi]\!]_{\mathcal{F}'}\} \\
&= \{w \in W \mid \exists U \in \{\{f(w,s_C,s_{\overline{C}}) \mid s_{\overline{C}} \in \Pi_{\overline{C}}^w\} \mid s_C \in \Pi_C^w\}. U \subseteq [\![\psi]\!]_{\mathcal{F}'}\} \\
&= \{w \in W \mid \exists U \in e(w,C). U \subseteq [\![\psi]\!]_{\mathcal{F}'}\} \\
&= [\![[C]\psi]\!]_{\mathcal{F}'},
\end{aligned}
$$

where the second equivalence holds by assumption. $\qquad\square$

3 Model Checking for the AMC

The model checking problem for the AMC consists in deciding, for state w in a model given either as a CGF or as an effectivity frame, and a closed AMC formula φ, whether $w \in [\![\varphi]\!]$. The problem is known to be in NP $\cap$ CO-NP for the AMC, and in PTime for ATL. A standard way to decide the problem is by reduction to parity games, enabling the use of recent quasipolynomial game-solving algorithms for model checking in the AMC.

Parity Games: A *parity game* $\mathcal{G} = (V, V_\exists, E, \Omega)$ is an infinite-duration two-player game, played by the players $\exists$ and $\forall$ (*Eloise* and *Abelard*). It consists of a set V of positions, with positions $V_\exists \subseteq V$ owned by $\exists$ and the others by $\forall$, a *move* relation $E \subseteq V \times V$, and a priority function $\Omega \colon V \to \mathbb{N}$. A *play* is a path in the directed graph (V, E) that is either infinite or ends in a node $v \in V$ with no outgoing moves. Finite plays $v_0 v_1 \ldots v_n$ are won by $\exists$ if and only if $v_n \in V_\forall$ (i.e. if $\forall$ is stuck); infinite plays are won by $\exists$ if and only if $\max\{p \mid \forall j \in \mathbb{N}.\, \exists k \geq j.\, \Omega(v_k) = p\}$ is even. A *(history-free) $\exists$-strategy* is a partial function $s\colon V_\exists \rightharpoonup V$ that assigns moves to $\exists$-nodes. A play *follows* a strategy s if for all $i \geq 0$ such that $v_i \in V_\exists$, $v_{i+1} = s(v_i)$. An $\exists$-strategy *wins* a node $v \in V$ if $\exists$ wins all plays that start at v and follow s.

We fix a closed formula φ for the remainder of this section and let $\mathsf{cl} = \mathsf{cl}(\varphi)$ denote its (Fischer-Ladner) *closure* [14].

Definition 2 (Model checking games, CGF semantics). *Given a model $\mathcal{M}$ consisting of a concurrent game frame $\mathcal{F} = (W, m, f)$ and a valuation ρ, the model checking game $\mathcal{G}^{\mathsf{CGF}}_{\mathcal{M},\varphi} = (V, E, \Omega)$ is the parity game defined by the following table, where game nodes $v \in V = V_\exists \cup V_\forall$ are of the shape $v = (w, \psi) \in W \times \mathsf{cl}$ or $v = (w, \psi, s_C) \in W \times \mathsf{cl} \times \Pi_C$; in the latter case, we require ψ to be a modality.*

node	owner	moves to	priority
$(w, \top)$	$\forall$	$\emptyset$	0
$(w, \bot)$	$\exists$	$\emptyset$	0
(w, p)	$\exists$	$\{(w, p) \mid w \in \rho(p)\}$	0
$(w, \neg p)$	$\forall$	$\{(w, \neg p) \mid w \in \rho(p)\}$	1
$(w, \varphi \wedge \psi)$	$\forall$	$\{(w, \varphi), (w, \psi)\}$	0
$(w, \varphi \vee \psi)$	$\exists$	$\{(w, \varphi), (w, \psi)\}$	0
$(w, \eta X.\,\psi)$	$\exists$	$\{(w, \psi[\eta X.\,\psi / X])\}$	$\mathsf{ad}(\eta X.\,\psi)$
$(w, [C]\psi)$	$\exists$	$\{(w, [C]\psi, s_C) \mid s_C \in \Pi_C(w)\}$	0
$(w, \langle C \rangle \psi)$	$\forall$	$\{(w, \langle C \rangle \psi, s_C) \mid s_C \in \Pi_C(w)\}$	0
$(w, [C]\psi, s_C)$	$\forall$	$\{(f(w, s_C, s_{\overline{C}}), \psi) \mid s_{\overline{C}} \in \Pi_{\overline{C}}(w)\}$	0
$(w, \langle C \rangle \psi, s_C)$	$\exists$	$\{(f(w, s_C, s_{\overline{C}}), \psi) \mid s_{\overline{C}} \in \Pi_{\overline{C}}(w)\}$	0

Thus $\mathcal{G}^{\mathsf{CGF}}_{\mathcal{M},\varphi}$ is a parity game with at most $|W| \times |\mathsf{cl}| \times (|\Pi| + 1)$ nodes and at most $\mathsf{ad}(\varphi)$ priorities. We point out that the model checking game has winning (resp. losing) self loops at nodes (w, p) (resp. $(w, \neg p)$) such that $w \in \rho(p)$.

Lemma 2. *Let $\mathcal{F}$ be a concurrent game frame with set W of states, and let $w \in W$ be a state. Then we have $w \in [\![\varphi]\!]_{\mathcal{F}}$ if and only if the existential player wins the node (w, φ) in $\mathcal{G}^{\mathsf{CGF}}_{\mathcal{M},\varphi}$.*

Definition 3 (Model checking games, EF semantics). If the model $\mathcal{M}$ is given as an effectivity frame $\mathcal{F} = (W, e)$, the model checking game $\mathcal{G}^{\mathsf{EF}}_{\mathcal{M},\varphi} = (V = V_\exists \cup V_\forall, E, \Omega)$ is defined in the same way as for concurrent game frames, but with nodes $v = (w, \psi, U) \in W \times \mathsf{cl} \times \mathcal{P}(W)$ replacing the nodes $(w, \psi, s) \in W \times \mathsf{cl} \times \Pi$ (with ψ of the form $[C]\psi'$ or $\langle C \rangle \psi'$). The modal moves are given by the table below; all other moves, and also node ownership and the priority assignments, are as in the CGF model checking game given above.

node	owner	moves to	priority
$(w, [C]\psi)$	$\exists$	$\{(w, [C]\psi, U) \mid U \in e(w, C)\}$	0
$(w, \langle C \rangle \psi)$	$\forall$	$\{(w, \langle C \rangle \psi, U) \mid U \in e(w, C)\}$	0
$(w, [C]\psi, U)$	$\forall$	$\{(v, \psi) \mid v \in U\}$	0
$(w, \langle C \rangle \psi, U)$	$\exists$	$\{(v, \psi) \mid v \in U\}$	0

In this case, $\mathcal{G}^{\mathsf{EF}}_{\mathcal{M}, \varphi}$ is a parity game with at most $|W| \times |\mathsf{cl}| \times (2^{|W|} + 1)$ nodes and at most $\mathsf{ad}(\varphi)$ priorities.

Corollary 1. *Let $\mathcal{F}$ be an effectivity frame with set W of states, and let $w \in W$ be a state. Then we have $w \in [\![\varphi]\!]_{\mathcal{F}}$ if and only if the existential player wins the node (w, φ) in $\mathcal{G}^{\mathsf{EF}}_{\mathcal{M}, \varphi}$.*

We point out that the above reductions to parity games make all transitions in models explicit, leading to games with a relatively large number of game nodes. Next, we present an alternative solution method that directly evaluates (sub)formulas over the state space W. This method does not explicitly construct the model checking games, and thereby avoids the blowup in state space incurred by the reduction to parity games.

Definition 4 (One-step evaluation). *Given sets $V \subseteq W$ and $\mathbf{X} = X_0, \ldots, X_k \subseteq V \times \mathsf{cl}$, we define a monotone function $\mathsf{prop}_V : \mathcal{P}(V \times \mathsf{cl})^{k+1} \to \mathcal{P}(V \times \mathsf{cl})$ evaluating propositional operators over argument sets $\mathbf{X} = (X_0, \ldots, X_k)$ by*

$$\mathsf{prop}_V(\mathbf{X}) = V \times \{\top\} \cup \{(w, p) \mid w \in \rho(p)\} \cup \{(w, \neg p) \mid w \notin \rho(p)\} \cup$$
$$\{(w, \varphi \wedge \psi) \mid \{(w, \varphi), (w, \psi)\} \subseteq X_0\} \cup$$
$$\{(w, \varphi \vee \psi) \mid \{(w, \varphi), (w, \psi)\} \cap X_0 \neq \emptyset\} \cup$$
$$\{(w, \eta X. \psi) \mid (w, \psi[\eta X. \psi / X]) \in X_{\mathsf{ad}(\eta X. \psi)}\}$$

Building on this, we define functions $f^{\mathsf{cgf}}_V : \mathcal{P}(V \times \mathsf{cl})^{k+1} \to \mathcal{P}(V \times \mathsf{cl})$ and $f^{\mathsf{ef}}_V : \mathcal{P}(V \times \mathsf{cl})^{k+1} \to \mathcal{P}(V \times \mathsf{cl})$ that evaluate formulas step by step using CGF semantics and EF semantics, respectively; formally, we put

$$f^{\mathsf{CGF}}_V(\mathbf{X}) = \mathsf{prop}_V(\mathbf{X}) \cup \{(w, [C]\psi) \mid \exists s_C \in \Pi^w_C. \forall s_{\overline{C}} \in \Pi^w_{\overline{C}}. (f(w, s_C, s_{\overline{C}}), \psi) \in X_0\}$$
$$\cup \{(w, \langle C \rangle \psi) \mid \forall s_C \in \Pi^w_C. \exists s_{\overline{C}} \in \Pi^w_{\overline{C}}. (f(w, s_C, s_{\overline{C}}), \psi) \in X_0\}$$
$$f^{\mathsf{EF}}_V(\mathbf{X}) = \mathsf{prop}_V(\mathbf{X}) \cup \{(w, [C]\psi) \mid \exists U \in e(w, C). (U \times \{\psi\}) \subseteq X_0\} \cup$$
$$\{(w, \langle C \rangle \psi) \mid \forall U \in e(w, C). (U \times \{\psi\}) \cap X_0 \neq \emptyset\}$$

Definition 5. *Given $V \subseteq W$ and a monotone function $f : \mathcal{P}(V \times \mathsf{cl})^{k+1} \to \mathcal{P}(V) \times \mathsf{cl}$, we define the nested fixpoints*

$$\mathsf{E}_f = \eta_k X_k. \eta_{k-1} X_{k-1}. \ldots . \nu X_0. f(X_0, \ldots, X_k)$$
$$\mathsf{A}_f = \overline{\eta_k} X_k. \overline{\eta_{k-1}} X_{k-1}. \ldots . \mu X_0. \overline{f}(X_0, \ldots, X_k)$$

where $\eta_i = \nu$ for even i and $\eta_i = \mu$ for odd i; also, $\overline{\nu} = \mu$ and $\overline{\mu} = \nu$. We use $\overline{f}$ to denote the dual function to f. These functions evaluate parity objectives over the functions f and $\overline{f}$, intuitively solving the associated model checking game by inlining intermediate game nodes that arise from the necessity to evaluate modalities; crucially, the domain of these fixpoint computations is $\mathcal{P}(V) \times \mathsf{cl}$ rather than the full set of nodes from the model checking game.

Lemma 3 ([11]). *Let $\mathcal{F}$ be a CGF with set W of states, and let $w \in W$ be a state. Then $w \in \llbracket \varphi \rrbracket_{\mathcal{F}}$ if and only if $(w, \varphi) \in \mathsf{E}_{f_W^{\mathsf{CGF}}}$; if $\mathcal{F}$ instead is an effectivity frame, then we have $w \in \llbracket \varphi \rrbracket_{\mathcal{F}}$ if and only if $(w, \varphi) \in \mathsf{E}_{f_W^{\mathsf{EF}}}$.*

The above implies a model checking algorithm that computes the nested fixpoints in Definition 5 iteratively. This algorithm can be extended to support *local* model checking, i.e. may avoid exploring the entire model if satisfaction of the target formula in a target state can be determined early [11].

4 Implementation

4.1 COOL The Coalgebraic Ontology Logic Solver

The model checking implementation is set within the framework provided by COOL/COOL 2/COOL-MC - the COalgebraic Ontology Logic solver - a coalgebraic reasoner and model checker for modal fixpoint logics [8–10], implemented in OCaml. We contribute implementations of the translation from CGFs to effectivity frames and instantiations of the generic tool to both CGFs and effectivity frames. In the COOL model checking framework, models are represented as functions $f \colon W \to FW$ where F is a set constructor obtained from the grammar

$$F, G {::} = M \mid id \mid \mathcal{P}F \mid F \times G \mid M \to G$$

where M ranges over sets. The application FX of a set constructor to a set X is defined as follows: $MX = M$, $id\,X = X$, $\mathcal{P}FX = \mathcal{P}(FX)$, $(F \times G)\,X = FX \times GX$, $(M \to G)\,X = M \to GX$, where $\mathcal{P}(FX)$ is the power set of FX and $M \to GX$ is the set of functions with domain M and codomain GX. Functions of type $W \to FW$ are referred to as *F-coalgebras* [23] but we refrain from delving into the general theory. Effectivity frames are then *F*-coalgebras for $F = \mathcal{P}\mathsf{At} \times (\mathcal{P}(\mathsf{Ag}) \to \mathcal{P}\mathcal{P})$, while concurrent game frames are *F*-coalgebras for $F = \mathcal{P}\mathsf{At} \times \mathbb{N}^{\mathsf{Ag}} \times (\Pi \to id)$.

As the datatype to represent all types of models, COOL defines the algebraic datatype `functor_element`.

4.2 Parity Game Model Checking

In COOL-MC, the generation of parity games is implemented as a higher order function that traverses the input model and formula and translates all connectives into game nodes as described in Sect. 3, implementing the interpretation of modal operators using a function it receives as an argument. Each occurrence of a coalitional modality generates a region consisting of inner nodes; the remaining nodes are called outer nodes, and form a further region. The structure of inner nodes differs depending on the semantics. Each region has an initial node (either the occurrence of a modality or the root node of the game). The type for game nodes consists of one algebraic type encoding the structure of inner nodes, and a further algebraic type encoding whether a node is an inner node or not.

The parity game thus created then can be solved using any parity game solver; COOL-MC, which we employ for the experiments in this work, uses the implementation of Zielonka's algorithm that is provided by PGSolver [24].

4.3   Local Model Checking

Within the COOL-MC framework, we also implement the alternative local model checking algorithm described in Sect. 3, again realized as a higher order function that receives the semantic function for modalities as an argument, as in Definition 4, and then computes the relevant fixpoints by Kleene approximation; indeed, both coalitional modalities $[C]$ and $\langle C \rangle$ are implemented within the same function. Recall that the semantics of $[C]\psi$ is given by

$$\exists s_C \in \Pi_C(w).\, \forall s_{\overline{C}} \in \Pi_{\overline{C}}(w).\, (f(w, s_C, s_{\overline{C}}), \psi) \in X_0$$

in CGF semantics (using notation introduced in Sect. 2). The quantifiers in this formula are implemented by traversing all joint moves of C and $\overline{C}$, respectively.

The implementation of the semantics of coalitional modalities for local model checking over CGFs (see Listing 1) enumerates all possible moves using the recursive function `try_out` until it either has found a witnessing move for the coalition, or it has proved that there is not such a move. This is achieved by advancing the move of the coalition if the reached world with the `curr_move` does not satisfy the argument formula (`x` not contained in `argset`). This correspond the existential search for the move of the coalition in the concurrent game structure semantics definition. When the current move of the coalition reaches a satisfying world then instead the move of the opposition `d_bar` is advanced using `next_move`. The recursion stops if the individual moves have hit the `move_bounds` so that all relevant moves have been explored.

For comparison, in EF semantics, the semantics of $[C]\psi$ is realized by implementing the function

$$\exists U \in e(w, C).\, (U \times \{\psi\}) \subseteq X_0.$$

Listing 2 shows the new implementation in COOL. Due to the generic representation of the models and formulas, a few lines of code are needed to extract the `coalition` and `eff_func` from the model and formula representation. Then the `capabilities` of the coalition corresponding to the $e(w, C)$ in the semantics definition are looked up (`H.find`) in the hash table underlying the effectivity function. The subsequent `exists` search then corresponds to picking the set U in the semantics, and checking containment of U in `argset` corresponds to the clause $U \subseteq [\![\varphi]\!]_\sigma$ in the semantics definition.

```
let coalition_modal_pred _ form xi_of_x argset argset_dual =
  let open Output in
  let* (box, d) =
    match form.HCFml.node with
    | F.HCENFORCES (d, _) -> return (true, d)
    | F.HCALLOWS (d, _) -> return (false, d)
    | _ -> Error (InvalidInput ( "coalition_modal_pred"))
  in
  let argset = if box then argset else argset_dual in
  begin (* xi_of_x = (moves_1, ..., moves_n, result_fun) *)
    let* tuple = M.fe_tuple_to_list xi_of_x in
    let n = ((length tuple) - 1) in (* number of agents *)
    ...
    let rec try_out curr_move =
      let x = Hashtbl.find result_fun curr_move in
      if mem x argset then
        begin
          let next = next_move_for move_bounds d_bar curr_move in
          if next = curr_move then return true else try_out next
        end
      else
        begin
          let next = next_move_for move_bounds d curr_move in
          if next = curr_move
          then return false
          else try_out (reset d_bar next)
        end
    in
    (* start exploring with move where everyone chooses 1 *)
    try_out (List.init n (fun _ -> 1) )
  end ||> fun x -> if box then x else not x
```

Listing 1. Semantics of the coalitional modalities for local model checking in concurrent game structure semantics

4.4   Converting CGFs to EFs

To calculate the induced effectivity function at a given state in a CGF, we implement the translation procedure described in Definition 1. As the set of states stays the same, the overall conversion is just a matter of converting the structure at each state individually. We define a function `effectivity_function_of_game_form`, which we iterate over all states to compute the resulting effectivity function (see Listing 3).

As the effectivity function assigns an effectivity set to every coalition, the implementation creates an empty hash table and then iterates over the set `all_coalitions` and calls the internal function `enforced_sets_for_coalition` to calculate the effectivity. This is done by calculating all admissible joint moves of the given coalition. Then the set `all_moves` is read off from the domain

```
let coalition_modal_effectivity_pred _ form xi_of_x argset argset_dual =
  let open Output in
  let* (box, coalition) =
    match form.HCFml.node with
    | F.HCENFORCES (d, _) -> return (true, d)
    | F.HCALLOWS (d, _) -> return (false, d)
    | _ -> Error (InvalidInput ( "coalition_modal_effectivity_pred"))
  in
  let argset = if box then argset else argset_dual in
  begin
    let* eff_func = M.fe_function_to_htbl __LINE__ xi_of_x in
    let  capabilities = H.find eff_func (Set (map M.int coalition)) in
    ...
    return (exists (fun succs -> subset succs argset) capabilities)
  end ||> fun x -> if box then x else not x
```

Listing 2. Semantics of the coalitional modalities for local model checking in effectivity frame semantics

```
let effectivity_function_of_game_form game_form =
  match game_form with
  | Tuple (* (moves_1, ..., moves_n, result_func)*) ->
      ...
    let effectivity = H.create (number_of_agents) in
      ...
    let move_is_extension grand_move coal coal_move =
      for_all2 (fun x y -> (nth grand_move (x-1)) = y) coal coal_move
    in
    let result_states coal coal_move = (* reached with partial move *)
      let result_state_for_extension grand_move =
        if move_is_extension grand_move coal coal_move
        then Some (result_func grand_move) else None
      in
      Set (filter_map result_state_for_extension all_moves)
    in
    let moves_for_coalition coal =
      let actions_for_coal = map (fun x ->(*{1,...,moves_x}*)) coal in
      all_combinations actions_for_coal
    in
    let enforced_sets_for_coalition coal =
      Set (map (result_states coal) (moves_for_coalition coal))
    in
    iter (fun coal ->
        H.add effectivity coal (enforced_sets_for_coalition coal)
      ) all_coalitions; Function effectivity
```

Listing 3. Conversion from CGFs to effectivity frames

of `result_func` and filtered for each of the partial moves of the coalition to get all grand moves extending the partial move of the coalition. Mapping the `result_func` over the resulting set gives the set of states reached by the joint move. After the effectivity functions have been computed, our conversion procedure filters the effectivity functions to only contain minimal sets. This operation ensures minimal runtime and space usage in all further processing of the resulting effectivity frames.

5   Experiments

We conduct various experiments, evaluating the performance of the different implementations of model checking for the alternating-time μ-calculus in comparison with each other, and (on the ATL fragment) in comparison to the state-of-the-art ATL model checker MCMAS. In more detail, we run benchmarks on the following implementations.

- Our implementation of concurrent game frame semantics model checking as an instance of COOL. We denote these algorithms by CGF_g (model checking by parity game reduction, using the implementation of Zielonka's algorithm provided by PGSolver [24] to solve the resulting games) and CGF_l (local model checking), respectively.
- Our implementation of model checking for effectivity frame semantics as an instance of COOL, writing EF_g (model checking by parity game reduction) and EF_l (local model checking), respectively, to denote the pure model checking algorithms; the combined algorithms that first perform the transformation from CGF to EF and subsequently model check are denoted by EFC_g and EFC_l.
- The symbolic multi-agent system model checker MCMAS (version 1.3.0) [17]; in contrast to the implementations within COOL, MCMAS handles additional epistemic modalities, and uses ordered binary decision diagram (OBDD) representations of models and formulas. On the other hand, MCMAS does not support the full AMC, so that the comparison between our implementations in COOL-MC and MCMAS is restricted to ATL.

In benchmarks, we use hyperfine [21] to average the measured values over at least five executions and set a timeout of 200 s. All experiments have been executed on a machine with an AMD Ryzen 7 2700 CPU and 32GB of RAM. An artifact containing the source code of our implementations, evaluation scripts and benchmarking sets for all experiments described below will be made available.

5.1   Random Formulas over Random Games

In a first experiment, we evaluate our implementations within COOL on random AMC formulas of increasing size on random concurrent game frames, each with 10 states, 4 agents, and either 2 or 10 moves per agent (due to the differences in model representation, a meaningful comparison with MCMAS on random models does not appear feasible).

5.2   Castle Game

The *castle game* has been used for benchmarking in previous work on ATL model checking [13, 22]. The game is parametrized over the number of castles and the health points all castles start with. Each castle has a corresponding knight that can, in each turn, either be sent out to attack another castle or stay and defend the castle. In each turn, all knights decide concurrently which other castle they want to attack or if they want to stay at their castle and defend. A knight who has attacked in one turn needs to stay and rest in the next turn. A castle that has its knight defending it or resting can block one attack. Each unblocked attack on a castle reduces that castle's number of health points by one. When no health points are left, the castle has lost the game and can no longer attack; this situation is indicated by propositional atoms lost_a, where a is a knight.

For the castle game we check the following AMC formulas (which are expressible in ATL) for satisfaction in the initial state. The formula $\nu X. \neg\mathsf{lost}_a \wedge [\{a\}]X$ expresses that the knight a has a strategy ensuring that her castle never gets destroyed. We check this formula for each $a \in \mathsf{Ag}$. Moreover, the formula

$$\mu X. \left(\left(\textstyle\bigwedge_{a \in C} \neg\mathsf{lost}_a\right) \wedge \left(\textstyle\bigwedge_{a \in \mathsf{Ag} \setminus C} \mathsf{lost}_a\right)\right) \vee [C]X$$

expresses that the coalition C has a joint strategy to ensure that all other castles are eventually destroyed while none of the allied castles (belonging to C) are destroyed. We check this formula for one coalition of each size.

The castle game has the property that almost none of the joint moves are equivalent, i.e. almost all joint moves lead to a different outcome. The specification in MCMAS uses separated local states of the agents.

The encoding of the castle game with n castles and h health points in COOL uses $(2 \times H)^n$ as state space where $2 = \{t, f\}$ and $H = \{x \mid 0 \leq x \leq h\}$.

5.3   Modulo Game

The second game we consider is constructed to showcase the benefit of effectivity frames in cases with many equivalent joint moves. The *modulo game* is parameterized over the number of agents, the number n of moves per agent, and a base number for the modulo calculation, which is also the number of states. Each state is uniquely identified by satisfaction of an atom p_i that witnesses that the current state corresponds to a sum of i. In each turn, the agents concurrently choose a number between one and n. The game then moves to the state corresponding to the sum of the played numbers plus the number of the previous state, modulo the base.

For the modulo game, we check the following formulas for satisfaction in the initial state (in which the mentioned sum starts at 0). The formula

$$\varphi_1 := \textstyle\bigwedge_{0 \leq i < \mathsf{base}} \mu X. \, p_i \vee [C]X$$

expresses that the coalition C has a joint strategy to reach any state eventually; this property can be expressed in ATL. On the other hand, the formula

$$\varphi_2 := \nu X. \, \mu Y. \, (X \wedge (p_0 \vee [C]Y) \wedge (p_{\mathsf{base}/2} \vee [C]Y))$$

expresses the Büchi property that coalition C has a strategy to ensure that a sum of 0 as well as a sum of base/2 occur infinitely often. We check both formulas for one coalition of each size. We use the first formula for comparison with MCMAS. The second formula requires the full AMC and is hence not covered by MCMAS, so it is used only for benchmarking the different COOL implementations. We fix 10 as the base throughout.

The (explicit-state) encoding of the modulo game in COOL is straightforward, having ten states and the expected transition function and valuation.

5.4 Results

The results of the experiment on *random formulas* are summarized in Fig. 2. Superscripts indicate the number of actions available to each of the agents; all models generated have 10 states. The vertical axis denotes the mean cumulative runtime including model conversion time for one run, i.e. an average over 25 instances of the model checking problem, each instance averaged over at least 5 runs. The horizontal axis denotes the size of formulas in terms of the number of connectives.

As indicated in the theoretical discussion of the individual model checking algorithms, inference using EF starts with a runtime offset owed to the conversion from CGF models to EF model. At small problem sizes, this overhead makes EF-based model checking slower than direct model checking on CGFs; however, the runtime eventually breaks even, with the EF-based algorithm outscaling the CGF implementation. In our benchmark parameters, this happens at a formula size near 16 for the local variants, and for much larger formulas for the game based implementations.

The *conversion times* for the modulo game are shown in Fig. 3. The effort required for converting from a concurrent game frame to an effectivity frame grows with both the number of states in the model and the number of moves. The modulo game always has exactly 10 states, and increasing the number of moves per agent leads to a blowup in the individual transition functions. The transition function has an entry for every joint move, so its domain has size $\mathsf{moves\_per\_agent}^{|Ag|}$, leading to the drastic growth visible in Fig. 3. The figure also shows the expected increase in conversion time when the number of agents is increased.

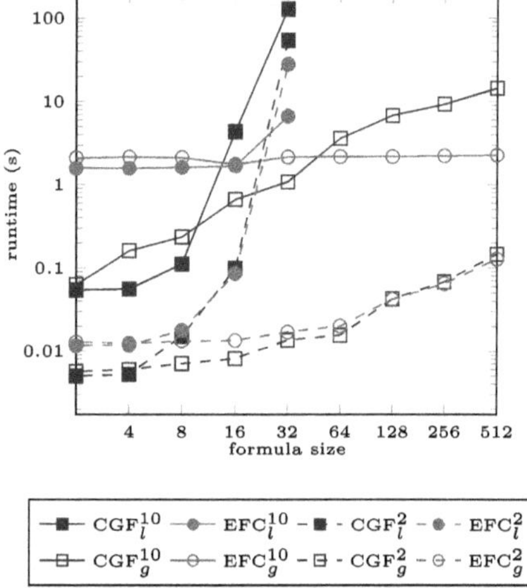

Fig. 2. Model checking random formulas.

number of agents is increased. The behaviour for the castles game is similar but scaled up due to the larger state space.

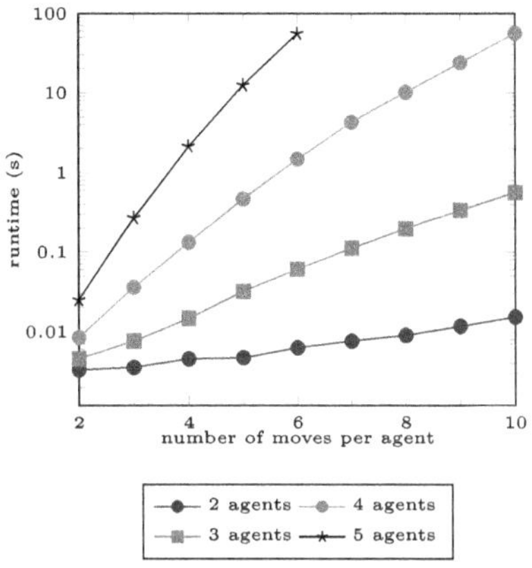

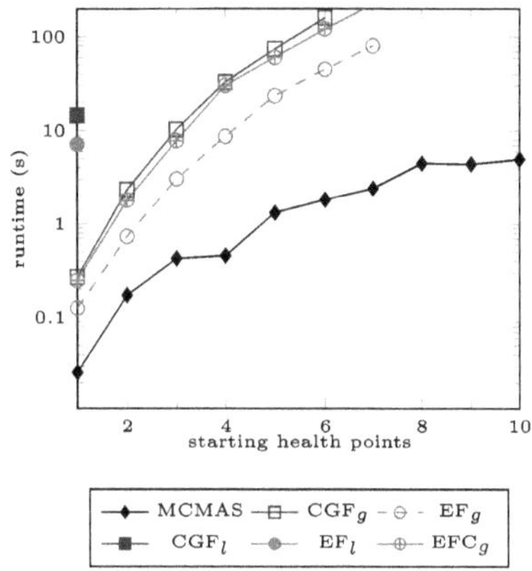

Fig. 3. Modulo game conversion time.   **Fig. 4.** Castle game runtime (4 castles).

The average runtime results for evaluating all described formulas over the *castles game* models are depicted in Fig. 4; they appear to show that MCMAS consistently outperforms our implementations in COOL. Both tools have similar performance characteristics with respect to increasing the starting health points affecting the number of states in the game and also the number of turns necessary to defeat a castle. The local variants of the COOL model checking perform significantly worse than the game-based variants on both the CGF and the EF semantics, possibly due to better scaling of Zielonka's algorithm as provided by PGSolver in comparison with the naive fixpoint approximation implemented in COOL. It is also notable here that while the model checking on effectivity frames generally performs better – as seen best in Fig. 4 – the cost of conversion outweighs this speedup at this model size. If a large number of formulas is to be evaluated on the same model, the cost of conversion is expected to amortize

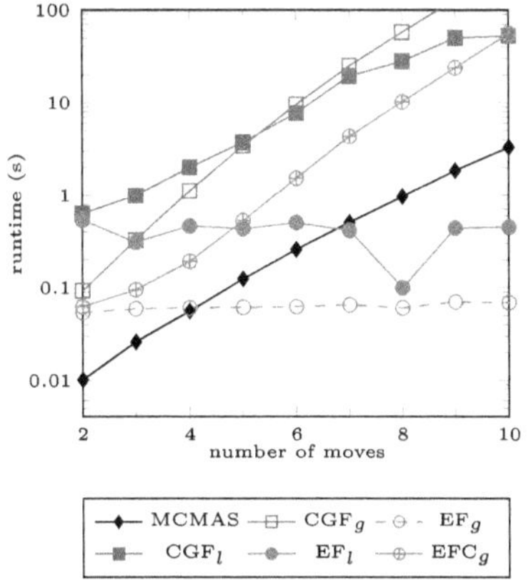

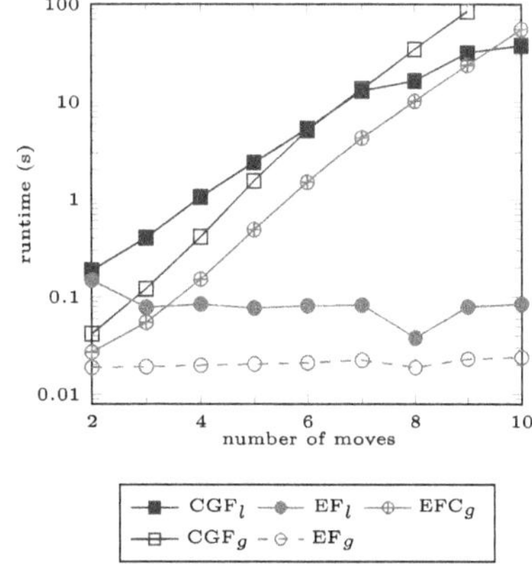

Fig. 5. Modulo game (φ_1, 4 agents).   **Fig. 6.** Modulo game (φ_2, 4 agents).

due to the slight speedup and might make this approach faster in the end. One reason why effectivity frames do not result in more speedup in the castle game is presumably that almost no moves are equivalent, so the resulting EFs are comparatively large.

On the other hand, EF-based model checking performs very well when model checking the *modulo games* against the formula φ_1 (Fig. 5). The runtime of the EF-based semantics is essentially constant, while the runtime under CGF semantics increases with the number of moves per agent. The constant runtime under EF semantics is as expected, since any coalition has a constant number of non-equivalent joint moves. The effectivity function hence scales mainly with the number of agents and only up to a fixed limit w.r.t. the number of moves per agent or the size of the coalition. The two EF implementations in COOL even outperform MCMAS with increasing number of moves and agents. Note however, that the conversion time increases sharply with increasing model size (similar to Fig. 3), nivellating to some extent the performance gain when the system is not initially specified as effectivity frame. Spikes in the plots appear to be due to symmetry in the model structure. In Fig. 6, the four COOL implementations are compared on the modulo game with the Büchi property benchmark formula φ_2. As expected, the effectivity-frame-based implementations have roughly constant performance while the concurrent game frame implementations slow down drastically with increased number of moves per agent.

The three benchmarking sets we evaluate all vary different parameters. Generally, effectivity frame semantics plays out its advantages when there are many equivalent joint moves of a coalition, as showcased by the constant performance in the modulo game (e.g. Fig. 5). Compared to model checking random formulas on concurrent game frames directly, there is a speedup for large input formulas as seen in Fig. 2. The BDD-based approach of MCMAS outperforms COOL on examples with few equivalent joint moves (see Fig. 4). On smaller problems, the speedup of the EF semantics is diminished by the initial cost of conversion when the problem is not specified as effectivity frame directly (see Fig. 3). We note however that COOL is the only model checker able to handle the full AMC where the speedup is potentially multiplied due to unfolding of the fixpoints increasing the effective formula size (see Fig. 6).

6   Conclusions and Future Work

We have analysed efficiency gains in model checking for the alternating-time μ-calculus afforded by converting concurrent game frames into effectivity frames in a preprocessing step. We have evaluated this method in comparison both with a direct implementation of model checking on concurrent game structures within the same overall framework and with the state-of-the-art ATL symbolic model checker MCMAS. Results show favourable performance of the preprocessing method on large systems that do not have succinct symbolic representations.

A large part of the computational cost of our method lies in the preprocessing, so future development will focus on optimizations of the conversion. In particular, since the conversion is entirely per-state, there is potential for parallelization. Moreover, further performance gains might be achievable by converting from concurrent game frames to effectivity functions by-need, that is, only when a modality is actually being evaluated on the state at hand.

References

1. Alur, R., et al.: JMOCHA: a model checking tool that exploits design structure. In: International Conference on Software Engineering, ICSE 2001, pp. 835–836. IEEE Computer Society (2001). https://doi.org/10.1109/ICSE.2001.919196
2. Alur, R., Henzinger, T.A., Kupferman, O.: Alternating-time temporal logic. J. ACM **49**, 672–713 (2002). https://doi.org/10.1145/585265.585270
3. Belardinelli, F., Jamroga, W., Kurpiewski, D., Malvone, V., Murano, A.: Strategy logic with simple goals: tractable reasoning about strategies. In: International Joint Conference on Artificial Intelligence, IJCAI 2019, pp. 88–94 (2019). https://doi.org/10.24963/IJCAI.2019/13
4. Čermák, P., Lomuscio, A., Mogavero, F., Murano, A.: MCMAS-SLK: a model checker for the verification of strategy logic specifications. In: Biere, A., Bloem, R. (eds.) CAV 2014. LNCS, vol. 8559, pp. 525–532. Springer, Cham (2014). https://doi.org/10.1007/978-3-319-08867-9_34
5. Chatterjee, K., Henzinger, T.A., Piterman, N.: Strategy logic. Inf. Comput. **208**(6), 677–693 (2010). https://doi.org/10.1016/j.ic.2009.07.004
6. Goranko, V., Jamroga, W.: Comparing semantics of logics for multi-agent systems. In: Information, Interaction and Agency, pp. 77–116 (2004). https://doi.org/10.1007/1-4020-4094-6_3
7. Goranko, V., Kuusisto, A., Rönnholm, R.: Game-theoretic semantics for ATL$^+$ with applications to model checking. Inf. Comput. **276**, 104554 (2021). https://doi.org/10.1016/j.ic.2020.104554
8. Gorín, D., Pattinson, D., Schröder, L., Widmann, F., Wißmann, T.: Cool – a generic reasoner for coalgebraic hybrid logics (system description). In: Demri, S., Kapur, D., Weidenbach, C. (eds.) IJCAR 2014. LNCS (LNAI), vol. 8562, pp. 396–402. Springer, Cham (2014). https://doi.org/10.1007/978-3-319-08587-6_31
9. Görlitz, O., Hausmann, D., Humml, M., Pattinson, D., Prucker, S., Schröder, L.: COOL 2 – a generic reasoner for modal fixpoint logics (system description). In: Automated Deduction, CADE 2023. LNCS, vol. 14132, pp. 234–247. Springer, Heidelberg (2023). https://doi.org/10.1007/978-3-031-38499-8_14
10. Hausmann, D., Humml, M., Prucker, S., Schröder, L., Strahlberger, A.: Generic model checking for modal fixpoint logics in COOL-MC. In: Verification, Model Checking, and Abstract Interpretation, VMCAI 2024. LNCS, vol. 14499, pp. 171–185. Springer, Heidelberg (2024). https://doi.org/10.1007/978-3-031-50524-9_8
11. Hausmann, D., Schröder, L.: Game-based local model checking for the coalgebraic μ-calculus. In: Concurrency Theory, CONCUR 2019. LIPIcs, vol. 140, pp. 35:1–35:16. Schloss Dagstuhl - Leibniz-Zentrum für Informatik (2019). https://doi.org/10.4230/LIPIcs.CONCUR.2019.35
12. van der Hoek, W., Lomuscio, A., Wooldridge, M.J.: On the complexity of practical ATL model checking. In: Autonomous Agents and Multiagent Systems, AAMAS 2006, pp. 201–208. ACM (2006). https://doi.org/10.1145/1160633.1160665
13. Kański, M., Niewiadomski, A., Kacprzak, M., Penczek, W., Nabiałek, W.: Unbounded model checking for ATL. Studia Informatica **25**(1–2) (2021). https://doi.org/10.34739/si.2021.25.01

14. Kozen, D.: Results on the propositional μ-calculus. Theor. Comput. Sci. **27**, 333–354 (1983). https://doi.org/10.1016/0304-3975(82)90125-6
15. Kupke, C., Marti, J., Venema, Y.: Size measures and alphabetic equivalence in the μ-calculus. In: Logic in Computer Science, LICS 2022, pp. 18:1–18:13. ACM (2022). https://doi.org/10.1145/3531130.3533339
16. Litak, T., Pattinson, D., Sano, K., Schröder, L.: Model theory and proof theory of coalgebraic predicate logic. Log. Methods Comput. Sci. **14**(1) (2018). https://doi.org/10.23638/LMCS-14(1:22)2018
17. Lomuscio, A., Qu, H., Raimondi, F.: MCMAS: an open-source model checker for the verification of multi-agent systems. Int. J. Softw. Tools Technol. Transfer **19**(1), 9–30 (2015). https://doi.org/10.1007/s10009-015-0378-x
18. McMillan, K.L.: Applying SAT methods in unbounded symbolic model checking. In: Brinksma, E., Larsen, K.G. (eds.) CAV 2002. LNCS, vol. 2404, pp. 250–264. Springer, Heidelberg (2002). https://doi.org/10.1007/3-540-45657-0_19
19. Mogavero, F., Murano, A., Perelli, G., Vardi, M.Y.: Reasoning about strategies: On the model-checking problem. ACM Trans. Comput. Log. **15**(4), 34:1–34:47 (2014). https://doi.org/10.1145/2631917
20. Pauly, M.: A modal logic for coalitional power in games. J. Log. Comput. **12**(1), 149–166 (2002). https://doi.org/10.1093/logcom/12.1.149
21. Peter, D.: Hyperfine (2023). https://github.com/sharkdp/hyperfine
22. Pilecki, J., Bednarczyk, M.A., Jamroga, W.: SMC: synthesis of uniform strategies and verification of strategic ability for multi-agent systems. J. Log. Comput. **27**(7), 1871–1895 (2017). https://doi.org/10.1093/logcom/exw032
23. Rutten, J.J.M.M.: Universal coalgebra: a theory of systems. Theor. Comput. Sci. **249**(1), 3–80 (2000). https://doi.org/10.1016/S0304-3975(00)00056-6
24. tcsprojects: PGSolver. https://github.com/tcsprojects/pgsolver

Locating Concurrency Errors in Windows .NET Applications by Fuzzing over Thread Schedules

Filip Kliber and Pavel Parízek[(✉)]

Charles University, Prague, Czech Republic
{kliber,parizek}@d3s.mff.cuni.cz

Abstract. We present a new fuzzing technique for multithreaded C# programs running on the .NET platform. It is built upon the .NET Profiling library, supported by CLR (Common Language Runtime) on Windows, and uses configurable strategies for the fuzzing process. During execution of the subject program, the fuzzing algorithm controls thread scheduling and preemption through suspending and resuming threads at specific code locations that we call stop points. For the purpose of driving the fuzzing process, we have designed a hybrid systematic-random strategy that gradually finds yet unexplored thread schedules. Results of experiments with programs from the SCT benchmark collection show that our tool is able to find errors triggered by specific thread interleavings, and within practical time limits.

1 Introduction

Fuzzing has become a very popular approach to discovering inputs and configurations that may trigger runtime errors in software systems [8, 26]. It has been successfully used to detect critical security vulnerabilities. The basic idea of automated fuzzing is to execute the subject program repeatedly many times, each time with a different input, and monitor its behavior and output for errors. Usually, the first input can be selected randomly, so that an arbitrary execution is tested, and then other subsequent inputs are derived by the fuzzer algorithm with the goal of exploring alternative execution traces (e.g., different control-flow paths). The main practical benefit of fuzzing is the ability to generate especially inputs that represent corner cases not considered and expected by human developers.

A specific category of programs are those involving multiple concurrent threads of execution. It is well known that threads may interleave in many ways, under the control of a system scheduler, and some thread schedules (interleavings) may trigger a concurrency error state, such as deadlock or atomicity violation (data race), or assertion violations caused by inconsistent concurrent data updates. Concurrency errors are, in general, very hard to find, because they are exposed only by specific rare interleavings of threads' execution [14].

Various approaches to detecting runtime concurrency errors, and related bugs in the program source code, have already been proposed, including those based on model checking [7, 19] with partial order reduction [1], reduction to sequential programs and subsequent verification [11, 12], dynamic analysis [5, 18, 25], systematic concurrency

G. Ernst and K. Y. Rozier (Eds.): SPIN 2025, LNCS 15945, pp. 125–141, 2026.
https://doi.org/10.1007/978-3-032-06847-7_7

testing [4,6,13,21], and static program analysis [15,16]. Many techniques and tools in each category have been developed so far. But all of these approaches have their well-known limitations and challenges, for example state explosion in the case of model checking, practical scalability and performance in the case of systematic concurrency testing, and (im)precision caused by abstraction in the case of techniques based on static analysis. For any realistic large and complex program (software system), there is really a huge number of possible ways in which execution of program threads can interleave, so then it is very hard to find some of the rare interleavings that may actually trigger some concurrency error at runtime.

Therefore it is no surprise that fuzzing is yet another promising approach for detecting concurrency-related bugs and vulnerabilities in multithreaded programs [2,9,10,24]. With this perspective of fuzzing applied to multithreaded programs, a thread schedule may be considered as the subject program's runtime configuration or specific input too. A scheduler in the operating system (e.g., Windows kernel) or virtual machine (such as .NET CLR) then represents a part of environment that influences the program's execution and outcome.

We have created a new fuzzing tool for multithreaded C# programs to be executed with the .NET platform runtime (CLR) on Windows systems. The tool uses a custom runtime analysis, built upon the .NET profiling library [27], together with a component responsible for driving the exploration of thread schedules. It exercises the behavior of subject programs under many different thread schedules, but with fuzzer-controlled thread preemption only at statements and code locations (method calls) configured by the user. Our main contribution is a new algorithm that gradually computes the set of distinct thread schedules to be explored, with the goals of achieving high coverage and trying to avoid repeated execution of the same thread interleaving multiple times. In each iteration of the main fuzzing loop, the algorithm yields the specific thread schedule to be followed. The key characteristic of our algorithm is that it navigates the fuzzer towards thread schedules that were not covered already, doing that with sufficient precision and reliability.

In this paper, we present (1) an overview of the fuzzing tool and the main algorithms it uses, (2) a list of related technical challenges that we have faced and our solutions, including the discussion of key design choices and general insights, and (3) experimental evaluation together with discussion of our experience and lessons learned.

The paper has the following structure. We provide an overview of the proposed fuzzing approach in Sect. 2, and technical details about main components of the fuzzing tool in Sects. 3 and 4, respectively. Specifically, the main algorithms used by the fuzzer are described in Sects. 2–4. Then we present results of experimental evaluation in Sect. 5, also discussing our observations. We compare our approach with related work in Sect. 6, and conclude with the discussion of high-level insights and possible directions of future work in Sect. 7.

2 Overview

First we describe the overall architecture of our fuzzing tool and introduce few key concepts. The tool consists of three main components:

- a runtime profiler that monitors execution of a subject program and watches for specific actions (e.g., calls of thread synchronization API methods),
- thread scheduling controller (manager) that is able to suspend and resume threads running within the .NET platform during execution of the program, and
- the actual fuzzing driver that implements our algorithm for exercising different thread schedules.

Input of the fuzzer tool includes (1) a binary program executable by the .NET runtime platform and (2) a user-defined set of program code locations where thread scheduling choices (preemption) should be triggered to cover interesting thread interleavings. These code locations are called *stop points* in our paper, and they are in particular calls of library procedures that implement synchronization operations (lock acquire, lock release, wait, notify, etc.) and calls of application methods. In the rest of this section, we present the whole fuzzing process and tasks performed by each component of the tool.

Algorithm 1 shows the key parts of our algorithm for exercising different thread schedules within the fuzzing process. The algorithm maintains a list of execution traces (thread schedules) that were already explored, and keeps running until it cannot find any new thread schedule that has not been covered yet. Details about the function that yields new thread schedules are provided in Sect. 4. In each iteration, the fuzzer picks one new thread schedule sch and executes the subject program under the schedule, with the runtime profiler attached. At the start of the whole fuzzing process, when the list of explored traces is empty, an arbitrary thread schedule is selected randomly.

Algorithm 1. Algorithm for exercising thread schedules

Input: The input binary program subject to fuzzing, represented by the symbol $binProg$, and the set of pre-configured stop points represented by the symbol $cfgStopPoints$.

$exploredTraces \leftarrow []$

while DRIVER::EXISTSUNEXPLOREDSCHEDULE($exploredTraces$) **do**

    $sch \leftarrow$ DRIVER::GETNEWTHREADSCHEDULE()

    EXECUTEWITHPROFILER($binProg, cfgStopPoints, sch$)

    $exploredTraces \leftarrow exploredTraces \cup sch$

end while

function ONREACHEDSTOPPOINT($curTh$)

    MANAGER::SUSPENDTHREADS($curTh$)

    $thsToEnable \leftarrow$ DRIVER::CHOOSETHREADSTORESUME(sch)

    MANAGER::RESUMETHREADS($thsToEnable$)

end function

The runtime profiler watches execution of the program, observing especially method calls, and creates a log of the whole trace that is reported to the user in case an error is detected. When the current active thread calls a method that corresponds to a pre-defined stop point, the following actions are performed by the fuzzer (see the function ONREACHEDSTOPPOINT in Algorithm 1):

1. Thread manager suspends ("freezes") some of the currently active threads, notably either just the current thread or all threads, depending on the configuration.
2. Fuzzing driver takes the set of suspended threads and from this set chooses, according to the schedule *sch* and the configured strategy, a subset that contains threads to become active at the given point.
3. Right after that, execution of all these selected threads is resumed ("thawed") by the thread manager.

Execution of the program then continues until another stop point is reached. Technical details about suspending and resuming threads are provided in Sect. 3.

When the execution of the subject program with profiling finishes, the top-level fuzzing algorithm continues with another iteration where an alternative thread schedule is considered. The explored schedules differ by the sets of threads suspended and resumed at individual stop points. Section 4 gives more details about: selection of threads to become runnable (active), strategies that we designed, and the procedure for generating thread schedules (traces) not yet explored.

In the rest of this paper, we illustrate specific aspects of the whole process on the example program in Fig. 1. It consists of two threads communicating through a shared buffer. The function `ComputeData` just performs some computation that produces a new value. Labels with the prefix `op_` are used below to identify the respective method call statements, which correspond to stop points.

```
interface Buffer() {                static void Main() {
  Add(int v);                         th1 = new Thread(thread1);
  int Remove();                       th2 = new Thread(thread2);
}                                     th1.start(); th2.start();
                                    }

thread1() {                         thread2() {
  op_11:                              op_21:
    int v = ComputeData();              int r = buffer.Remove();
  op_12:                              op_22:
    buffer.Add(v);                      Console.WriteLine(r);
}                                   }
```

Fig. 1. Example program with two threads

If threads in the program execute concurrently, there are six possible ways they can interleave, displayed in Fig. 2. These six interleavings correspond to thread schedules that should be explored to achieve full coverage.

The next two sections provide details about steps of the whole fuzzing process. In particular, we discuss our technical design decisions and challenges that we faced.

3   Runtime Profiler and Thread Manager

The runtime profiler component uses the low-level Microsoft Profiling API [27] for .NET applications, which is provided by the .NET runtime platform, to monitor program execution and observe interesting events. However, this is why our fuzzing tool

op_11	op_11	op_11	op_21	op_21	op_21
op_21	op_12	op_21	op_11	op_22	op_11
op_22	op_21	op_12	op_22	op_11	op_12
op_12	op_22	op_22	op_12	op_12	op_22

Fig. 2. All possible interleavings of threads in the example program

targets just applications running on Windows systems, because the Profiling API is supported only in releases of both .NET Framework and .NET Core on the Windows platform. When the fuzzing process is about to start, the profiler is attached to the subject program. During execution of the program, it receives notifications about relevant events from the .NET Common Language Runtime (CLR).

We have implemented and configured our profiler component such that it gets notified about calls of methods that represent possible stop points. In addition, the profiler is notified about certain system-level thread synchronization- and concurrency-related events, such as thread creation, and monitors the dynamic stack trace for each thread.

The thread manager, as already indicated, controls the actual interleaving of individual threads during program execution by the means of suspending and resuming threads at stop points. In this process, it tries to follow the given thread schedule, selected by the fuzzing driver component, as closely as possible.

One of the key design decisions behind our fuzzing algorithm was to enable *stop points* only at method call statements. This is, however, sufficient in practice because of the way the .NET platform works. All the relevant and interesting events from the perspective of concurrency fuzzing are represented by method calls—notably, accesses to properties (i.e., to object fields via getters/setters) and even basic thread synchronization operations (lock acquire, release, etc.). Users define stop points by the name of the method and its owner class. We assume that, in practice, users will mark as possible stop points especially the calls of application methods that may access data shared by multiple concurrent threads.

Originally we have considered stop points of two kinds, strong and weak. The difference is that only the current active thread is suspended at a weak stop point, while all runnable application threads are suspended at a *strong stop point*. Nevertheless, based on our initial experiments, we have found that strong stop points are superior, meaning that usage of weak stop points does not have any benefits with respect to coverage of thread interleavings. Therefore, from now on, by the term "stop point" we always refer to a strong stop point.

We had to define the procedure for suspending all threads very carefully, in particular to avoid unsafe states caused by preempting some application threads while they are executing within the kernel space or system libraries. The problem is that, when one application thread reaches a strong stop point, other threads may be located anywhere in managed or unmanaged code, executing syscalls or native library procedures, manipulating shared kernel resources and holding the respective global locks (e.g., when allocating heap memory on Windows), and so on. We have considered and implemented two solutions that we describe here:

- One solution, that we claim to be safe, is to stop threads only at well-defined locations in the application code—specifically, at the boundaries of application methods. When some application thread reaches a stop point, thread manager asks all the other application threads to stop, by setting a flag introduced for this purpose. The flag is checked by the runtime profiler handlers for method entry and exit events. In practice, thread manager just has to wait until every application thread is out of the kernel space (or finishes a library call), and reaches the nearest entry or exit point in the code of some application method. To be more specific, when thread T is executing a library method or syscall invoked by the application method M, the manager waits until T reaches either (i) the entry of some other application method (including getters and setters for properties) called from M or (ii) exit from M, whatever happens first.
- Another option is to stop threads "by force" at their current locations, even if they reside in the kernel space or execute a system library procedure, and solve the problem on a different level. In particular, to avoid deadlocks in the Windows heap allocator (when one thread would be stopped inside the implementation of std::malloc while holding the global allocation lock), it is necessary to use a custom allocator that manipulates a fixed memory pool.

However, only the second solution is now enabled in the profiler and thread manager.

An important related technical detail is the usage of a special service thread within the profiler and the manager, a thread that performs the following steps repeatedly in an infinite loop: waits until some application threads are suspended, passes an updated list of all suspended threads to the fuzzing driver component, and then asks the driver to decide which of the suspended threads should be resumed. The action of resuming threads is therefore also triggered by the special service thread. Supported strategies for deciding which threads to resume are described in Sect. 4. It is necessary to use a separate service thread for this purpose, because the runtime profiler exists directly inside the CLR and has no thread allocated on its own. When some application thread triggers an event watched-for by the profiler, the corresponding event handler defined by our profiler is executed by that same application thread.

Note also that the fuzzing driver can select no threads (an empty set) to be resumed at a given stop point and moment of time. In that case, the service thread just continues and asks the driver again, periodically, every few milliseconds.

4   Fuzzing Driver: Exploring Thread Schedules

The fuzzing driver component of our tool has two main duties: (1) recording information about executed traces and (2) selecting thread schedules to be explored.

First we provide important details regarding the management and recording of traces. A trace is collected by the runtime profiler for every execution of the subject program. It is a list of thread scheduling decisions made at stop points, where each entry in the list contains the following information: an ID of the application thread that was suspended, code location of the stop point (method signature and owner class name), and IDs of all threads that were selected by the driver to be resumed. The driver stores the set of all traces (sequences of thread scheduling choices) already explored, and uses this knowledge in future iterations of the fuzzing process to navigate program execution towards alternative thread schedules.

We mentioned before that we designed the algorithm for selecting thread schedules with two goals in mind: (1) to explore as many unique thread schedules as possible within available resources, including the time budget, and (2) to avoid repeated exploration of already visited thread interleavings.

Control over schedules is realized through selection of threads to resume execution at stop points. During the first execution of the subject program in a fuzzing session, when there are no saved explored traces, the choice of threads is completely random. In the case of all other executions, the individual scheduling choices are driven by a configurable strategy for exploring many distinct traces. We support two automated strategies, *random* and *systematic*, and interactive user control too.

The random strategy selects the thread to be resumed from a list of all suspended threads completely at random. In addition, it can be configured to insert specific delays between the moments when threads are suspended and resumed, further affecting how the final schedule looks like. A consequence of random selection is that some already explored thread schedules may be visited repeatedly (multiple times).

We have also created the interactive user driver, which human developers can use through a command-line interface, mainly for debugging and error reproduction. Users can simply resume threads by providing their IDs on the command line. A standard interactive debugger, when attached to the application, can be used to inspect the program state and confirm hypotheses about root causes of the observed errors.

4.1 Systematic Strategy

The proposed systematic strategy for exploring thread schedules is designed around a special data structure, which we call a *scheduling graph*. It is created at the start of each iteration of the fuzzing process, and captures the set of thread schedules corresponding to all previously executed traces. Nodes of the scheduling graph represent stop points and edges represent taken scheduling decisions. Every node is labeled with an integer value that expresses the total number of possible *distinct* unexplored thread schedules that could be executed from the stop point associated with the node. These values are used to find not-yet-explored alternative thread schedules (traces). Edges are labeled with thread IDs and (sequences of) executed program statements.

We illustrate the concept of scheduling graphs on our example program (Fig. 1) and the list of possible thread schedules (Fig. 2). Assuming that each execution of the program by the fuzzer takes a different schedule, in the order presented in Fig. 2 from left to right, then Fig. 3 shows what the scheduling graphs created after the respective iterations (executions) of the fuzzing process would look like.

Having such a graph after some iteration of the fuzzing process, a new thread schedule to be explored in the next round is derived simply by depth-first traversal of the graph from the root node along a path over nodes with non-zero labels, representing thread scheduling choices with possibly unexplored options. The following cases are considered in each step of the traversal:

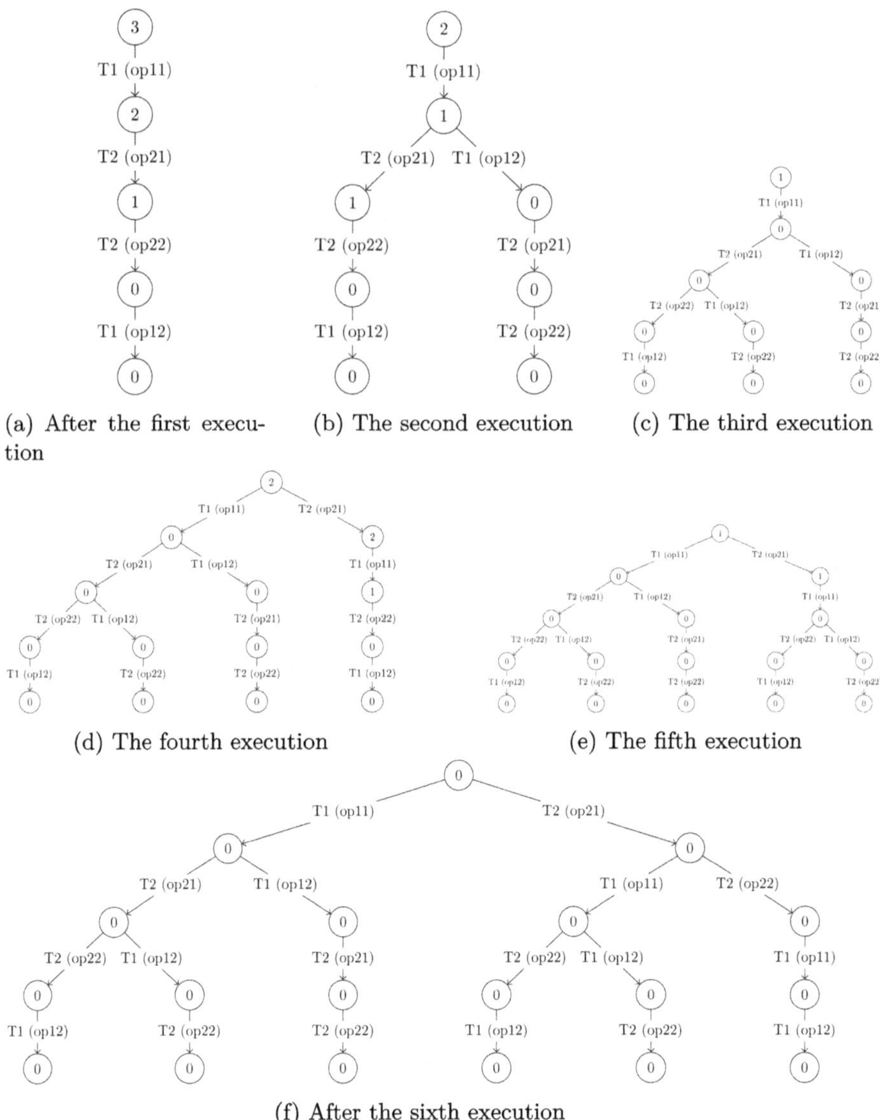

(a) After the first execution

(b) The second execution

(c) The third execution

(d) The fourth execution

(e) The fifth execution

(f) After the sixth execution

Fig. 3. Scheduling graphs for execution traces of the example program in Fig. 1

- If the current node is labeled with a non-zero value, and it has at least one child node with a non-zero value, then traversal proceeds by moving to one of the child nodes with a non-zero value attached.
- If the current node is labeled with a non-zero value, but all of its existing child nodes in the graph are labeled with zero, then the search produces a new thread schedule by choosing a thread that is not yet covered by any edge leading from the node. See,

---

Algorithm 2. Producing new thread schedules

Input: List of traces explored in previous iterations, $exploredTraces = [T_1, T_2, \ldots, T_n]$.

    function EXISTSUNEXPLOREDSCHEDULE($exploredTraces$)
        $graph \leftarrow$ CREATEGRAPH($explordTraces$)
        return $graph.root.value > 0$
    end function

    function GETNEWTHREADSCHEDULE
        $schedule \leftarrow$ EXTENDSCHEDULEBYDFS($[\,]$, $graph.root$, $graph$)
        return $schedule$
    end function

    function EXTENDSCHEDULEBYDFS(sch, cn, G)
        if $cn.value > 0 \land \exists w$ s.t. $(cn, w) \in G.edges \land w.value > 0$ **then**
            $sch \leftarrow sch$ ++ $(cn, w).label$
            EXTENDSCHEDULEBYDFS(sch, w, G)
        end if
        if $cn.value > 0 \land \neg\exists(cn, w) \in G.edges$ s.t. $w.value > 0$ **then**
            $sch \leftarrow sch$ ++ th s.t. $\forall(cn, w) \in G.edges \cdot (cn, w).label \neq th$
        end if
        return sch
    end function

---

for example, the root node of a graph in Fig. 3c. Traversal ends in this case, and the new thread schedule is returned.

- If the current node is labeled with zero, traversal backtracks.

Specifically, if the root node of the whole graph is labeled with zero, all thread schedules were explored and the fuzzing process terminates. The procedure for deriving new thread schedules is defined more formally in Algorithm 2. The helper function EXTENDSCHEDULEBYDFS attempts to extend the current schedule sch with thread choices represented by edges leading from the given node cn.

The procedure to create a scheduling graph for a set of already explored traces (thread schedules) is formalized in Algorithm 3. It is a key part of our systematic driver strategy for the fuzzer. The helper function CREATEPATHFORTRACE extends the graph with a path that represents the given trace, and returns a leaf node on the newly created path in the scheduling graph. The helper function UPDATEVALUESFORNODES iteratively updates the attached value for each node on the new path, in the direction from a leaf node to the root.

4.2 Discussion and Remarks

Here we discuss important properties of the fuzzing approach, especially with the systematic driver strategy, that we observed when performing initial experiments during the development.

Algorithm 3. Construction of scheduling graphs

Input: List of traces collected from previous iterations.
Output: Scheduling graph $G = (nodes, edges)$.

function CREATEGRAPH($[T_1, T_2, \ldots, T_n]$)
    $root \leftarrow$ NEWNODE(), $root.value \leftarrow 0$
    $G \leftarrow (nodes \leftarrow root, edges \leftarrow \emptyset)$
    for $T_i \in T_1, \ldots, T_n$ **do**
        $ln \leftarrow$ CREATEPATHFORTRACE(G, T_i)
        UPDATEVALUESFORNODES(G, ln)
    end for
end function

function CREATEPATHFORTRACE(G, T_i)
    $v \leftarrow G.root$
    for $th \in T_i$ **do**
        if $\neg \exists w$ s.t. $(v, w) \in G.edges \wedge (v, w).label = th$ **then**
            $w \leftarrow$ NEWNODE(), $w.value \leftarrow 0$, $G.nodes \leftarrow G.nodes \cup \{w\}$
            $e \leftarrow (v, w)$, $e.label \leftarrow th$, $G.edges \leftarrow G.edges \cup \{e\}$
        end if
        $v \leftarrow w$
    end for
    return v
end function

function UPDATEVALUESFORNODES(G, v)
    repeat
        $\exists!(w, v) \in G.edges$
        $w.value \leftarrow 0$
        $X \leftarrow \{x \mid (w, x) \in G.edges\}$
        for $x \in X$ **do**
            $w.value \leftarrow w.value + x.value$
        end for
        $m \leftarrow max($COUNTENABLEDTHREADS$(x) \mid x \in X)$
        $w.value \leftarrow w.value + max(0, m - |X|)$
        $v \leftarrow w$
    until $v = G.root$
end function

A practical limitation of the presented approach to systematic exploration of thread schedules is that it requires stable thread IDs across program executions, to be able to compare traces explored in different iterations. Using system thread IDs, provided by the OS kernel or .NET CLR, is not possible, because they change in each execution of the subject program. In our prototype implementation of the fuzzer tool, we have decided to use abstract thread IDs, integer numbers starting from 1 and incremented for every created thread. However, this solution does not work for programs that use a thread pool, such as in the case of .NET Task Parallel Library, where it is not certain that the same thread from a pool will handle a specific Task object in each execution.

The behavior of our fuzzing algorithm in practice also greatly depends on the set of pre-defined code locations to be used as stop points. Consider this scenario. The driver strategy resumes a small subset of suspended threads at a particular stop point, for example just one thread, and no other stop point is hit during continuation of that single thread's execution. In that case, execution of the whole program may end in a deadlock state, when the single thread finishes its execution and other threads are suspended. Users need to be aware of the fact that many iterations (runs) of the fuzzer, on many thread schedules (traces), may finish by such a spurious deadlock.

We recommend, based on our experience so far, to define a sufficiently large set of code locations as possible stop points. The stop points can be viewed as breakpoints and preemption locations in executions of individual threads. A small set of defined stop locations may result in a relatively high ratio of pathological thread schedules.

5 Experiments

We have created a prototype implementation of the proposed fuzzing approach, described in previous sections, in an open source tool. Complete source code is available in the repository https://github.com/d3sformal/threadfuzzer-net.

We evaluated the practical usefulness of our tool based on these important criteria: its ability to find runtime errors (e.g., assertion violations, crashes and deadlocks) triggered by specific thread interleavings, whether it has good performance, and coverage of distinct thread schedules. In this section, we describe the setup of our experiments and present the results of measurements, and then we dicuss main observations.

For the purpose of this evaluation, we performed experiments with programs from the SCTBench collection [23,28], which is quite a representative set of benchmarks regarding usage of concurrency. But first we had to translate all programs to the C# language, and modify some of the benchmark programs by refactoring class fields into class properties with automatically generated **get** and **set** methods. This was needed because our tool currently supports stop points only at method calls. Most of the programs in SCTBench are small, containing just few methods, so our configuration of stop point locations for each benchmark contains all of its applications methods, together with methods of the library class **SemaphoreSlim**.

The fuzzing tool can be, in practice, applied only to a standalone executable program, e.g. with the **Main** method, or to a partial program (library) with test cases. Additionally, it makes sense to use the fuzzer with programs that contain some checks for runtime error states during their execution. We reflect this in our evaluation by running the fuzzer on existing (unit) concurrency tests.

To properly analyze the benefits and limitations of our fuzzing tool, we ran experiments for three different configurations of the fuzzer, and for plain executions (without the profiler attached) as the baseline. In case of the baseline configuration, thread scheduling is fully controlled by the OS and .NET CLR. The other configurations are the following:

- Usage of the random strategy in the fuzzing driver component, denoted by the term *Random Fuzzing* in the table with data.
- Systematic strategy for the driver in the variant that always selects the first unexplored thread schedule found in the graph by DFS, denoted as *Systematic First* in the table with results.
- Systematic driver strategy combined with random selection of an unexplored thread schedule from all that could be found in the scheduling graph (*Systematic Random*).

All experiments were done on a personal computer running Windows 10 Pro 64-bit (10.0, Build 19045) with processor AMD Ryzen 7 PRO 5850U and 16 GB of memory. We have provided all that is needed to run experiments (scripts, benchmark programs, etc.) in the public repository with our implementation.

Table 1 presents the results of our experiments for all benchmarks and configurations. In each experiment, i.e. for each pair of configuration and benchmark, we ran our fuzzing tool with a hard upper limit of 150 iterations, meaning that at most 150 thread schedules were explored. The actual number of explored unique thread schedules varies just very slightly over the set of all runs; is in the range 145–150 for almost all experiments. We also set a time limit for each iteration of the fuzzing algorithm, that means for a single execution of the subject program with profiler, to 20 s. Values of these metrics are reported in the table for each benchmark and experiment:

- DoC, Degree of Concurrency, which is a statically approximated number of OS threads running during program execution.
- Lines of code, LoC, and size of the program code in kB.
- The percentage of runs that ended with an error detected by the fuzzer.
- Percentage of runs that timed out, typically after reaching a deadlock state where some applications threads are suspended and unable to finish their execution.
- Average execution time for all non-timedout iterations.

Data in Table 1 show that those configurations of the fuzzing tool (exploration strategies) that involve random choice have the greatest ability to find errors—indicated by the high percentages of runs where an error was detected and reported. Looking at results for these strategies, namely "Random Fuzzing" and "Systematic Random", we observe that percentages of explored thread schedules (within the bound of 150 runs) that trigger errors are comparable over the set of all benchmarks. The systematic strategy in the select-first variant discovers less error-triggering thread schedules, because it may reach the limit of 150 runs while exploring just one segment of the scheduling graph (and thus executing many thread schedules with a common prefix).

We also note that the percentage of timed out runs (iterations of the fuzzing algorithm) across the set of all experiments is very low. It means that just a small number of thread schedules ended in a deadlock state.

On the other hand, there is one notable exception to the general observed pattern—results obtained for the benchmark "TwoStage*100", which has an extremely high degree of concurrency. The scheduling graph is really big in this case, so that traversal of the graph causes the fuzzer to run out of the time limit.

Table 1. Results for all configurations and benchmarks

Benchmark				Baseline	Random Fuzzing	Systematic First	Systematic Random
Name	DoC	LoC (kB)	Metrics				
Account	4	51 (1.2)	Violated	2.7 %	28.7 %	14.0 %	30.0 %
			Timed out	0.0 %	0.0 %	4.0 %	1.3 %
			Time/Iter	33.8 ms	210.3 ms	233.6 ms	214.5 ms
BluetoothDriver	2	74 (1.7)	Violated	0.0 %	6.0 %	0.0 %	1.3 %
			Timed out	0.0 %	0.0 %	0.0 %	0.0 %
			Time/Iter	33.9 ms	550.8 ms	676.2 ms	532.1 ms
Carter01	5	67 (1.3)	Violated	0.0 %	0.0 %	0.0 %	0.0 %
			Timed out	0.0 %	12.0 %	7.3 %	20.7 %
			Time/Iter	29.8 ms	213.6 ms	213.2 ms	200.4 ms
CircularBuffer	3	86 (2.2)	Violated	0.0 %	59.3 %	6.0 %	69.3 %
			Timed out	0.0 %	0.0 %	1.3 %	1.3 %
			Time/Iter	29.7 ms	476.6 ms	226.5 ms	558.3 ms
Deadlock01	3	42 (1.0)	Violated	0.0 %	0.0 %	0.0 %	0.0 %
			Timed out	0.7 %	2.7 %	8.0 %	5.3 %
			Time/Iter	28.5 ms	201.3 ms	202.2 ms	196.1 ms
Lazy01	4	40 (0.9)	Violated	0.7 %	66.0 %	48.0 %	59.3 %
			Timed out	0.0 %	0.0 %	0.7 %	2.3 %
			Time/Iter	32.3 ms	571.3 ms	466.0 ms	538.5 ms
Queue	3	99 (2.6)	Violated	98.7 %	100.0 %	84.7 %	98.7 %
			Timed out	0.0 %	0.0 %	2.0 %	1.3 %
			Time/Iter	48.5 ms	763.0 ms	290.5 ms	748.6 ms
Reorder3	17	83 (2.3)	Violated	0.0 %	6.0 %	0.0 %	8.0 %
			Timed out	0.0 %	0.0 %	0.0 %	1.3 %
			Time/Iter	30.9 ms	464.4 ms	596.6 ms	473.4 ms
Reorder10	52	83 (2.3)	Violated	0.0 %	3.3 %	7.3 %	2.7 %
			Timed out	0.0 %	0.0 %	7.3 %	3.3 %
			Time/Iter	29.8 ms	3416.5 ms	5840.3 ms	3431.6 ms
tack	3	73 (1.6)	Violated	1.3 %	61.3 %	0.0 %	44.7 %
			Timed out	0.0 %	0.0 %	0.0 %	0.7 %
			Time/Iter	34.9 ms	848.0 ms	821.3 ms	828.3 ms
TokenRing	5	57 (1.3)	Violated	8.7 %	10.7 %	9.3 %	15.3 %
			Timed out	0.0 %	0.0 %	2.7 %	0.0 %
			Time/Iter	29.5 ms	208.8 ms	202.7 ms	210.7 ms
TwoStage100	101	71 (1.9)	Violated	0.0 %	0.0 %	0.0 %	0.0 %
			Timed out	0.0 %	100.0 %	9.3 %	100.0 %
			Time/Iter	30.0 ms	N/A	17032.8 ms	N/A
TwoStageSmall	3	89 (2.3)	Violated	0.0 %	5.3 %	0.0 %	0.7 %
			Timed out	0.0 %	0.0 %	2.0 %	0.7 %
			Time/Iter	29.0 ms	284.5 ms	310.7 ms	283.5 ms
WrongLock	9	61 (1.7)	Violated	0.0 %	13.3 %	13.3 %	20.7 %
			Timed out	0.0 %	0.0 %	8.7 %	4.0 %
			Time/Iter	29.8 ms	211.4 ms	218.4 ms	223.8 ms

6   Related Work

Lot of research has already been done on various techniques for detecting bugs and errors related to concurrency in software, including also techniques for fuzzing multithreaded programs that were published recently. Here we compare our approach to several closely related techniques with similar goals and ideas.

A popular way of detecting concurrency bugs is through systematic unit testing. The Coyote tool [4] provides infrastructure for writing and running concurrency unit tests for .NET. It focuses, in particular, on the task-based asynchronous programming model that is very popular in the context of C#, and therefore has a limited ability to detect low-level data races. Unlike our approach, which uses .NET Profiling library to get notifications about relevant events and to control thread scheduling, Coyote instruments the subject binary program with code that enables the Coyote engine to control execution and thread scheduling.

Search for concurrency errors driven by random choice is another popular approach, implemented by many techniques and tools. For example, Sen [21] has proposed an algorithm that uses random sampling to cover the whole space of partial orders (thread schedules) more uniformly, and later proposed a technique for detecting low-level races where the random testing (selection) of thread schedules to be explored in directed by information about possible data races that is provided by dynamic analysis [22]. The key idea of [22] is the following. Simple random testing is performed most of the time, with one exception—when the next statement to be executed in the current thread belongs to some candidate pair of racy statements, execution of that statement is delayed until the other statement from the pair is about the be executed in some other thread, in this way checking whether the two statements in a candidate pair can really be executed concurrently. The follow-up work by Park and Sen [17] focuses on detecting higher-level atomicity violations through combination of dynamic analysis with random search and delaying of lock operations. However, all these techniques detect concurrency errors (deadlocks, races, etc.), while our approach supports detection of general errors (violated assertions, crashes, failed tests, etc.) triggered by specific thread interleavings.

Many other algorithms for search over the space of possible thread schedules, in the context of systematic concurrency testing, have been published so far. Thomson et al. [23] present results of an empirical study that compares the performance of selected algorithms, including the following: basic depth-first search, preemption bounding, delay bounding, and controlled random scheduler. Our fuzzing tool supports most of these algorithms, either directly or through user configuration. The basic depth-first search corresponds to our systematic (first) strategy. Preemption bounding can be simulated through configuring the stop points in a specific way. Delay bounding then corresponds to usage of weak stop points combined with a timeout (bound) on resuming threads. Finally, the controlled random scheduler algorithm corresponds to our systematic random strategy with a strong stop point enabled at every method call statement.

Table 2 contains the results of additional experiments that we performed for the purpose of comparison with specific related techniques. We present data for the configuration of our fuzzer that corresponds to the delay-bounding algorithm. Data in both tables show that the configurations "systematic random" and "delay bounding" have roughly the same ability to find errors, measured as the percentage of runs that ended with an error detected by the fuzzer, but "systematic random" is significantly faster from the two. This also confirms an observation, reported in [23], that search strategies based on random choice perform surprisingly well. Another empirical study with similar observations has been published by Rungta and Mercer [20].

Table 2. Results for the delay-bounding algorithm

Benchmark	Delay Bounding		
	Violated	Timed out	Time/Iter
Account	22.7 %	0.0 %	1039.9 ms
BluetoothDriver	2.7 %	0.0 %	1108.9 ms
Carter01	0.0 %	94.0 %	1704.2 ms
CircularBuffer	68.0 %	0.0 %	2474.1 ms
Deadlock01	0.0 %	57.3 %	1110.8 ms
Lazy01	50.7 %	0.0 %	1227.4 ms
Queue	98.0 %	0.0 %	2640.3 ms
Reorder3	0.7 %	0.0 %	3123.7 ms
Reorder10	7.3 %	0.0 %	4882.4 ms
Stack	37.3 %	0.0 %	4219.3 ms
TokenRing	25.3 %	0.0 %	1237.9 ms
TwoStage100	0.0 %	100.0 %	n/a
TwoStageSmall	8.7 %	0.0 %	1213.8 ms
WrongLock	29.3 %	2.7 %	2751.8 ms

The fuzzing techniques that explicitly support concurrent software extend the well-established approaches by recording and using concurrency-related information for the purpose of exercising program behavior under different thread schedules, in particular to detect vulnerabilities triggered by concurrent execution of threads. But they are not dedicated primarily on detecting concurrency errors (deadlocks, races). Chen et al. [2] developed MUZZ, a fuzzing technique that computes program inputs with the goal of achieving higher coverage of program behaviors (execution paths) in the multithreaded context. It targets scenarios where the execution of specific control-flow paths in the code of individual threads depends on the input values, exercising both input-dependent and also thread interleaving-dependent paths. For this purpose, MUZZ performs thread-aware instrumentation of program code, with special handling of thread scheduling-related operations (e.g., lock, unlock, and fork), and uses the collected data to prioritize exploration of traces (executions) that correspond to previously-unseen thread interleavings. Jiang et al. [9] developed CONZZER, a framework for concurrency fuzzing that introduces two new features: (1) more precise coverage metric based on tracking runtime calling contexts, and (2) control of thread scheduling based on inserting breakpoints. Knowledge of runtime calling contexts is used by CONZZER to identify pairs of methods, whose concurrent execution should be tested. The breakpoints, inserted before-hand (statically) by code instrumentation, are used to enforce actual concurrent execution of specific method pairs through careful management of time delays at runtime. This idea of using breakpoints and injected delays is similar to our stop points, where threads are suspended and resumed. Wolff et al. [24] proposed another fuzzing technique for multithreaded programs, which avoids redundant exploration of similar thread interleavings through an approach inspired by partial order reduction used in model checking. The key idea is to explore just one thread interleaving from each set of thread interleavings that are equivalent in terms of observable sequences of memory access events and values read by invididual threads at specific code locations.

7   Conclusion

The proposed approach to fuzzing multithreaded programs achieves high coverage of thread schedules and has a reasonable performance, as the results of our evaluation show. In particular, it is able to find errors triggered by specific thread interleavings quite efficiently, within practical limits on time and memory. For all these reasons, we believe that it is useful in practice. Still, there is a large space for extensions, improvements of user experience, and optimizations. Here we outline several directions for future work.

In the current version of our prototype implementation, the set of stop point locations is defined manually. This could be, at least partially, automated with the help of static analysis of the application program code and usage of information provided by dynamic analysis running on-the-fly in scope of the fuzzing process. The SharpDetect tool [3] with appropriate plugins could be used for the dynamic analysis. Our prototype implementation could be extended with additional functionality too, such as with (1) the ability to fuzz-test programs compiled specifically for the AMD64 platform and (2) support for the very recent versions of the .NET platform.

While looking for available concurrency benchmarks, we have also noticed that there is no established benchmark suite of programs written in the C# language. Already published research papers refer to individual bugs from all sorts of different software packages. We believe that having a diverse collection of benchmark programs in C# for .NET would foster the future research. Such collection of benchmarks should contain also few large programs, to facilitate evaluation of scalability. In the specific case of our fuzzing tool, a challenge related to analyzing large programs with many threads would be the specification of a right set of stop points.

Acknowledgments. This work was partially supported by the Czech Science Foundation project 23-06506S and partially supported by the Charles University institutional funding project SVV 260698.

References

1. Abdulla, P., Aronis, S., Jonsson, B., Sagonas, K.: Optimal dynamic partial order reduction. In: Proceedings of POPL 2014. ACM (2014)
2. Chen, H., et al.: MUZZ: thread-aware grey-box fuzzing for effective bug hunting in multi-threaded programs. In: Proceedings of USENIX Security (2020)
3. Čižmárik, A., Parízek, P.: SharpDetect: dynamic analysis framework for C#/.NET programs. In: Deshmukh, J., Ničković, D. (eds.) RV 2020. LNCS, vol. 12399, pp. 298–309. Springer, Cham (2020). https://doi.org/10.1007/978-3-030-60508-7_16
4. Deligiannis, P., Senthilnathan, A., Nayyar, F., Lovett, C., Lal, A.: Industrial-strength controlled concurrency testing for C# programs with coyote. In Proceedings of TACAS 2023, LNCS 13994 (2023)

5. Flanagan, C., Freund, S.N.: FastTrack: efficient and precise dynamic race detection. In: Proceedings of PLDI 2009. ACM (2009)
6. Fonseca, P., Rodrigues, R., Brandenburg, B.: SKI: exposing kernel concurrency bugs through systematic schedule exploration. In: Proceedings of OSDI 2014, USENIX (2014)
7. Godefroid, P.: Software model checking: the VeriSoft approach. Formal Meth. Syst. Des. **26**(2) (2005)
8. Godefroid, P.: Fuzzing: hack, art, and science. Commun. ACM **63**(2) (2020)
9. Jiang, Z.-M., Bai, J.-J., Lu, K., Hu,S.-M.: Context-sensitive and directional concurrency fuzzing for data-race detection. In: Proceedings of NDSS 2022, The Internet Society (2022)
10. Ko, Y., Zhu, B., Kim, J.: Fuzzing with automatically controlled interleavings to detect concurrency bugs. J. Syst. Software **191** (2022)
11. Lal, A., Reps, T.: Reducing concurrent analysis under a context bound to sequential analysis. In: Gupta, A., Malik, S. (eds.) CAV 2008. LNCS, vol. 5123, pp. 37–51. Springer, Heidelberg (2008). https://doi.org/10.1007/978-3-540-70545-1_7
12. La Torre, S., Madhusudan, P., Parlato, G.: Reducing context-bounded concurrent reachability to sequential reachability. In: Bouajjani, A., Maler, O. (eds.) CAV 2009. LNCS, vol. 5643, pp. 477–492. Springer, Heidelberg (2009). https://doi.org/10.1007/978-3-642-02658-4_36
13. Musuvathi, M., Qadeer, S.: Iterative context bounding for systematic testing of multithreaded programs. In: Proceedings of PLDI 2007. ACM (2007)
14. Musuvathi, M., Qadeer, S., Ball, T., Basler, G., Nainar, P.A., Neamtiu, I.: Finding and reproducing heisenbugs in concurrent programs. In: Proceedings of OSDI 2008, USENIX (2008)
15. Naik, M., Aiken, A., Whaley, J.: Effective static race detection for Java. In: Proceedings of PLDI 2006. ACM (2006)
16. Naik, M., Park, C.-S., Sen, K., Gay, D.: Effective static deadlock detection. In: Proceedings of ICSE 2009. IEEE CS (2009)
17. Park, C.-S., Sen, K.: Randomized active atomicity violation detection in concurrent programs. In: Proceedings of FSE 2008. ACM (2008)
18. Park, S., Vuduc, R.W., Harrold, M.J.: Falcon: fault localization in concurrent programs. In: Proceedings of ICSE 2010. ACM (2010)
19. Rabinovitz, I., Grumberg, O.: Bounded model checking of concurrent programs. In: Etessami, K., Rajamani, S.K. (eds.) CAV 2005. LNCS, vol. 3576, pp. 82–97. Springer, Heidelberg (2005). https://doi.org/10.1007/11513988_9
20. Rungta, N., Mercer, E.: Clash of the titans: tools and techniques for hunting bugs in concurrent programs. In: Proceedings of PADTAD 2009. ACM (2009)
21. Sen, K.: Effective random testing of concurrent programs. In: Proceedings of ASE 2007. ACM (2007)
22. Sen, K.: Race directed random testing of concurrent programs. In: Proceedings of PLDI 2008. ACM (2008)
23. Thomson, P., Donaldson, A.F., Betts, A.: Concurrency testing using controlled schedulers: an empirical study. ACM Trans. Parallel Comput. **2**(4) (2016)
24. Wolff, D., Shi, Z., Duck, G.J., Mathur, U., Roychoudhury, A.: Greybox fuzzing for concurrency testing. In: Proceedings of ASPLOS 2024. ACM (2024)
25. Yoga, A., Nagarakatte, S., Gupta, A.: Parallel data race detection for task parallel programs with locks. In: Proceedings of FSE 2016. ACM (2016)
26. Zeller, A., Gopinath, R., Böhme, M., Fraser, G., Holler, C.: The Fuzzing Book. CISPA Helmholtz Center for Information Security. https://www.fuzzingbook.org/. Accessed Feb 2025
27. .NET Profiling API (library). https://learn.microsoft.com/en-us/dotnet/framework/unmanaged-api/profiling/. Accessed Feb 2025
28. SCTBench: A collection of benchmarks. https://github.com/mc-imperial/sctbench. Accessed Feb 2025

Resilience Through Automated Adaptive Configuration for Distribution and Replication

Scott D. Stoller[✉], Balaji Jayasankar, and Yanhong A. Liu

Department of Computer Science, Stony Brook University, Stony Brook, USA
{stoller,bjayasankar,liu}@cs.stonybrook.edu

Abstract. This paper presents a powerful automated framework for making complex systems resilient under failures, by optimized adaptive distribution and replication of interdependent software components across heterogeneous hardware components with widely varying capabilities. A *configuration* specifies how software is distributed and replicated: which software components to run on each computer, which software components to replicate, which replication protocols to use, etc. We present an algorithm that, given a system model and resilience requirements, (1) determines initial configurations of the system that are resilient, and (2) generates a reconfiguration policy that determines reconfiguration actions to execute in response to failures and recoveries. This model-finding algorithm is based on state-space exploration and incorporates powerful optimizations, including a quotient reduction based on a novel equivalence relation between states. We present experimental results from successfully applying a prototype implementation of our framework to a model of an autonomous driving system.

Keywords: automated configuration · resilience · fault-tolerance · replication · distributed systems · adaptive systems

1 Introduction

Increasingly sophisticated critical software systems are built from a myriad of interconnected components providing a multitude of functionalities and running on diverse hardware components. It is well-known that distributing and replicating functionalities can enhance reliability. Reliability can be further enhanced by making a system *resilient*, i.e., able to quickly adapt to failures in order to continue providing service. To achieve resilience with minimum cost, systems should be able to adapt to optimally utilize the available hardware resources after any sequence of hardware failures and recoveries. This paper presents a powerful automated framework for making complex systems resilient under failures, by

---

This material is based on work supported in part by ONR grants N000142112719 and N000142012751 and NSF grant CCF-1954837.

G. Ernst and K. Y. Rozier (Eds.): SPIN 2025, LNCS 15945, pp. 142–160, 2026.
https://doi.org/10.1007/978-3-032-06847-7_8

optimized adaptive distribution and replication of software components across heterogeneous hardware components with widely varying capabilities, ranging from compute servers to mobile devices. Key features of the framework, and the main contributions of this paper, are:

- *Precise system model*, based on key characteristics needed for automated optimized adaptive distribution and replication, covering software, hardware, replication protocols, and configuration, including dependence, compatibility, and capacity constraints. Identifying these characteristics, thereby enabling automation of these decisions for a large class of systems, is a key contribution. The model supports a variety of hardware device types, because the challenge we address is especially important in cyber-physical systems. A *configuration* specifies how software is distributed and replicated: which software components run on each computer, which software components are replicated, the number of replicas and replication protocol used for each, etc.
- *Resilience requirements* specifying the *failure model*—the types of failures, as well as their numbers and rates, that must be tolerated—and the *critical functionalities*—functionalities that must be continuously available. The limit on failure rate helps ensure that the system has the opportunity to reconfigure in response to some failures, before additional failures occur. A few illustrative scenarios appear in Sect. 5. We focus on hardware failures, because software failures are addressed using, e.g., retry blocks and N-version programming, not distribution and replication.
- *Configuration policy generator*, in the form of a model-finding algorithm that solves two core problems: given a system model and resilience requirements, (1) determine *initial configurations* (i.e., configurations before any failures occur) that are *resilient*, i.e., continuous availability of all critical functionalities is guaranteed if the system starts in that state and appropriate reconfiguration actions are taken after each failure and recovery; and (2) generate a *reconfiguration policy* that determines reconfiguration actions to execute in response to each failure and recovery. Reconfiguration actions include adding and removing software replicas; starting, stopping, and moving running software; and replacing running software with alternative software with the necessary functionalities but different hardware requirements, software dependencies, or QoS. Precise modeling of replication protocols ensures that the best one is used in each context.
- *Quotient reduction based on a novel equivalence relation between states*, and other optimizations, to increase scalability of state-space exploration performed by the configuration policy generator.
- *Prototype implementation and experimental results* from successfully applying it to a model of an autonomous driving system.

We implemented the entire framework, with explicit-state state-space exploration, in DistAlgo [6], an extension of Python with support for logic quantifications with patterns as queries (and for high-level distributed programming, which would be useful later to implement a runtime system for reconfiguration).

This allows the mathematical definitions of reconfiguration, resilience, etc. to be translated into code directly and easily.

We also implemented most of the framework using the SMT solver Z3 [8], with a configuration policy generator based on symbolic state-space exploration. We encountered multiple difficulties, including the need to use Boolean encodings of our models and constraints due to lack of direct support for quantification over sets and functions, lack of direct support for recursion, and limited support for optimization problems.

Many details, including further discussion of our Z3 implementation, are omitted due to space limitations and are available in [12].

2   Related Work

Many works on replication protocols refer to support for reconfiguration. This means that the protocol supports dynamically adding and removing replicas, not that it addresses the system configuration problems we tackle.

There is a sizable literature on self-adaptive systems, surveyed in [10,13,14]. Weyns [13] identifies four essential tasks for self-adaptation. Our work is primarily self-healing and self-configuration—plus some overlap with self-protection and self-optimization—in the context of distribution (placement) and replication of computations. Work on self-adaptive systems addresses these tasks, especially self-healing, mainly in the context of other aspects of system configuration, such as network routing. Self-adaptation of these other aspects of configuration is complementary to our work.

Several works address configuration of distribution and replication of computations in specific settings. For example, Daidone et al. [1] use stochastic simulations to evaluate how configuration parameters—e.g., heartbeat rate for failure detection—of a single service replicated on a homogeneous set of servers affect the system's performance. Machida et al. [7] optimize static placement of VMs providing independent services on servers with different capacities, to reduce the number of servers needed to tolerate a given number of crashes. Stoiescu et al. [11] design a software architecture that supports adding and removing software components that implement fault-tolerance mechanisms, such as replication and retry blocks, at runtime. HAMRAZ [5] is a novel programming language whose compiler decides where to replicate objects and method invocations to ensure specified availability (and other) properties in distributed systems with multiple administrative domains and subject to Byzantine attacks.

Our work has two novel features compared to existing work. First, our definition and analysis of resilience are recursive. This enables more precise analysis of resilience, compared with traditional non-recursive definitions, for systems that reconfigure in response to failures, because it checks, for each sequence of failures, whether the system configuration created in response to those specific failures can tolerate exactly the additional failures that can occur—based on the assumptions in the failure scenario—in that failure state. Furthermore, our analysis synthesizes configurations and reconfiguration policies and provides precise resilience guarantees for them in advance, rather than simply making a "best effort" in each state without advance guarantees. Second, we consider automatic selection of replication protocols. Jointly optimizing the choice of replication protocol and the number and placement of replicas for each software component, and making them adaptive (varying over time), enables our method to achieve higher resilience with given resources than prior work. Also, our work uses a more comprehensive system model than much prior work, taking software dependencies, hardware compatibility, resource constraints, etc., into account. *We have not found any work in the literature on self-adaptive systems or fault-tolerance that has these two features or can generate guaranteed-resilient (with respect to a failure model) adaptive replication policies like those in the examples in Sect. 8.*

3   System Model

Our object-oriented system model captures characteristics of hardware components, software components, and replication protocols that affect adaptive configuration for resilience, including dependence, compatibility, capacity, and distribution constraints.

3.1   Hardware

Class *Computer* is used to model devices that can run application software: embedded computers, laptops, cell phones, etc. It has attributes specifying the: OS and CPU architecture; computational resources, namely, cores and RAM; networking capabilities (wired, wifi, and/or cellular); power sources, which may be internal (e.g., battery) or external; integrated devices, represented as a set *devices* of types of integrated devices, e.g., GPS or camera.

We model power sources, because a power source can be a single point of failure for multiple devices.

Class *Device* is used to model stand-alone devices, i.e., devices not integrated into computers. It has attributes specifying the: device type *type*, e.g., camera; and power sources (none for devices (e.g., power sources) that do not require a power source).

Stand-alone devices are assumed to be network-connected and accessible by all computers.

Network hardware (e.g., routers and wireless access points) is not currently included in our model; we plan to add it.

3.2   Software

Our model considers dependencies between software components as well as dependencies on hardware. This is especially useful for systems in which larger software components have been decomposed into smaller ones, which can be distributed and replicated differently to make the best use of limited resources.

The model captures characteristics that determine whether a software component must be replicated to achieve resilience or whether starting a fresh instance of the software after a failure is sufficient.

The *Software* class has attributes specifying the: functionality *fn*, e.g., map database or planning; set *fnReq* of required functionalities, i.e., functionalities provided by other software and used by this software; set *devices* of types of required devices, e.g., camera for vision software; required OS and CPU architecture, if any; resource requirements, namely, cores and RAM; and required networking capabilities (wired, wifi, and/or cellular).

It also has the following Boolean attributes specialized to resilience analysis:

- *deterministic*: whether the software is deterministic (i.e., different instances that receive the same sequence of client requests produce the same outputs). This affects which replication protocols can be used.
- *migratable*: whether a running instance of the software can be moved to another computer. This depends on capabilities of the computing platform (e.g., the OS) as well as characteristics of the software: support for migration of essential volatile state (if any), redirection of network connections (if any), etc.
- *persisState*: whether the software maintains essential mutable persistent state.
- *preferred*: whether this software is preferred, e.g., provides higher QoS than non-preferred software with the same functionality. This supports generating reconfiguration policies that achieve graceful degradation of QoS due to failures.
- *remoteUse*: whether the software can be used remotely by software components running on other computers.
- *resumable*: whether the software functions correctly after the computer it is running on fails and then recovers. This holds if the software does not have essential volatile state and can resume functioning correctly if it misses events while it is down.
- *singleInstance*: whether there should be at most one (possibly replicated) instance of this software in a system. This generally holds if it maintains data that should be unique system-wide. For example, a vehicle's computer vision system might run multiple independent object recognition components (one per camera) and feed the results into a single-instance scene understanding component that creates an integrated model of the vehicle's surroundings.
- *slowStarting*: whether the software's startup time exceeds the acceptable downtime of its functionality during reconfiguration. If so, replication, rather than starting new instances, must be used to achieve continuous availability of this software.
- *smallPersisState*: whether the persistent state (if any) is small, in the sense that the time needed to copy it over the network during reconfiguration is acceptable.

3.3 Knowledge Base for Replication Protocols

Our model includes a knowledge base for replication protocols, capturing key characteristics needed to support automated selection of the best applicable protocol in each context. The *RepProt* class has attributes: *sync*, a Boolean indicating whether the protocol is only for synchronous systems; *failTypes*, the types of handled failures; *active*, a Boolean indicating whether active or passive replication is used; *progressQ*, an enum with allowed values majority and all, indicating the size of a quorum needed for successful execution of a client request; and *reconfigQ*, an enum with allowed values majority and one, indicating the size of a quorum needed for successful reconfiguration, a.k.a. membership change. For brevity, we omit generic attributes of software components, such as required OS; they can trivially be added.

In active replication, all replicas receive and process requests from clients; this is suitable for deterministic applications only. In passive replication, one replica, called the *active replica* or *primary*, receives and processes requests from clients and disseminates the resulting state changes and outputs to the other replicas.

For traditional primary-backup protocols and chain replication protocols, *progressQ* = all and *reconfigQ* = one. For asynchronous state-machine replication protocols, such as Paxos [3,4] and Raft [9], typically *progressQ* = majority and *reconfigQ* = majority.

A recovered replica can resume participating in a replication protocol without reconfiguration, provided it was not removed from the membership while it was down.

3.4 Systems and Configurations

The *System* class has attributes specifying the sets *hw* and *sw* of hardware and software components, respectively, and a Boolean *sync* indicating whether the system is synchronous. A *configuration* of a system specifies which software components run on each computing device, which software components to replicate, and, for each replicated software component, which replication protocol to use to coordinate the replicas. The inlined UML class diagram shows classes *Config*, *SwInst* (for unreplicated software instances), *RepSwInst* (for replicated software instances), etc. Multiplicity symbols 1, ?, and * mean "one", "optional", and "multiple" (zero or more), respectively. For example, the outgoing edges of *RepSwInst* show that its attributes include *computers* (a set of computers where replicas run) and *primary* (an optional primary replica). For an instance *rsi* of *RepSwInst*, we sometimes treat *rsi.repProt.processQ* and *rsi.repProt.reconfigQ* as numbers; for example, if *rsi.repProt.processQ* = majority, then *rsi.repProt.processQ* is treated as $\lceil (|rsi.computers| + 1)/2 \rceil$.

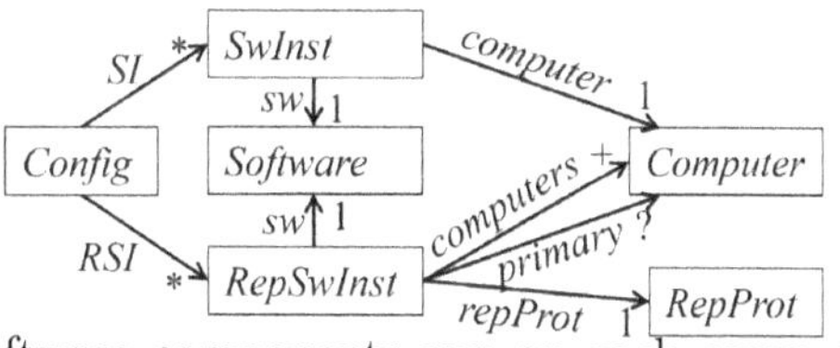

Our model captures several types of constraints on configurations. A configuration *cfg* is *valid* if it satisfies these constraints. *Dependence constraints* include dependence of software on software—for example, a software component that requires a functionality provided by another—and of software on hardware—for example, a software component that requires data from a GPS. *Compatibility constraints* include hardware/software compatibility—for example, between binary executables and CPU architectures such as ARM and x86, between a system's synchrony guarantee and a replication protocol's synchrony requirements—as well as software/software compatibility—for example, between application software and OS. *Capacity constraints* ensure software resource requirements do not exceed hardware resource capacities, currently enforced for total number of CPU cores and amount of memory needed by the software components running on each computing device; capacity constraints for other resources, e.g., network throughput, can easily be added. *Distribution constraints* require co-location of software components—specifically, if a software component depends on functionality of another software component accessible only through a local API—and ensure there is at most one instance of each single-instance software component.

4   Failure Model

A *failure model* specifies the types of hardware failures considered, which types of hardware are subject to each type of failure, and limits on the number and rate of failures. We currently consider only crash failures and recovery from crashes. We assume detection of crash failures is complete, i.e., every failed component is eventually suspected of failure. In synchronous systems, we further assume that it is accurate, i.e., correct components are not suspected of failure.

A *failure* is modeled with class *Failure*. It has attributes specifying the hardware component hw that failed and the failure type *type*, currently only crash. A *recovery* from a failure is modeled in the same way as a failure; it represents the end of that failure.

A *failed set* is the set of failures representing the currently failed hardware components with their failure types, i.e., the failures that have occurred minus the recoveries. A *state* of a system is a pair $\langle cfg, fs \rangle$, where *cfg* is a configuration, and *fs* is a failed set.

A hardware component h is *live* with failed set *fs*, denoted $live(h, fs)$, if h is not in *fs*, and, if h requires an external power supply, at least one of its power supplies is live. Functionality *fn* is *available* on computer c in configuration *cfg* with failed set *fs*, denoted $availOn(fn,c,cfg,fs)$, if: either (1) the configuration contains a software instance *si* running on a live computer that (1a) *si.sw* has functionality *fn*, (1b) if $\neg si.sw.remoteUse$, then $c = si.computer$, (1c) for each functionality *fn'* required by *si.sw*, $availOn(fn',c,cfg,fs)$, and (1d) each device type d required by *si.sw* is available on c, i.e., c has an integrated device of type d or there is a live network-accessible device of type d; or (2) the configuration contains a replicated software instance *rsi*, and there exists a subset Q of *rsi.computers*, such that (2a) *rsi.sw* has functionality *fn*, (2b) $|Q| \geq$ *rsi.repProt.progressQ* and $\forall c \in Q.live(c, fs)$, (2c) if $\neg rsi.sw.remoteUse$ then $c \in Q$, (2d) if *rsi.repProt.active*, then for each computer c' in Q, $availOn(si.sw.fnReq, c', cfg, fs)$ holds and each device type d required by *si.sw* is available on c, and (2e) if $\neg rsi.repProt.active$, then $availOn(si.sw.fnReq, rsi.primary, cfg, fs)$ holds and each device type d required by *si.sw* is available on *rsi.primary*. Functionality *fn* is *available* in state $\langle cfg, fs \rangle$, if it is available on some computer.

We define *removeDead(cfg, fs)* to be *cfg* with dead software instances on failed computers removed. A software instance on a failed computer is *dead* if its execution cannot usefully be resumed if the failed computer recovers; its execution can usefully be resumed if the software is resumable, is not slow-starting, and has persistent state.

A *failure model* characterizes the sequences of failures a system is expected to tolerate. A failure model limits (1) the maximum numbers of failed components, for various failure types and hardware types, and (2) the rate of failures, abstracted as the maximum number of "simultaneous" failures, i.e., the maximum number of failures that can occur in a time interval shorter than the system's reconfiguration time.

The *FailBound* class is used to specify limits on failures of a given type. It has fields *hwType* (a hardware type), *fType* (a failure type), n (the maximum number of hardware components of *hwType* that can have a failure of type *fType* in any state) and *maxSimult* (the maximum number of simultaneous failures of type *fType* for hardware components of type *hwType*, or None if there is no limit, other than n).

The *FailureModel* class has fields *fBounds* (a set of instances of *FailBound*) and *maxSimult* (the maximum number of simultaneous failures across all failure types, or None if there is no limit beyond the one implied by *fBounds*).

These bounds could be based on a probabilistic failure model and probabilistic reliability requirement, e.g., the expected time until a combination of failures that exceeds these bounds is at least k years. The bound on simultaneous failures should also take common-cause failures into account.

A failed set *fs* is *consistent* with failure model *fm*, denoted *consistent(fs, fm)*, if it satisfies the bounds on the total number of failures and the numbers of failures of each type.

Given a failed set *fs* and a failure model *fm*, let *nextFS(fm,fs)* be the set of failed sets that are consistent with *fm*, are strict supersets of *fs*, and are reachable from *fs* by one set of simultaneous failures . Let *nextFSworst(fm,fs)* contain the subset-maximal members of *nextFS(fm,fs)*. When a maximal set of simultaneous failures occurs in a state with failed set *fs*, one of these "worst-case" failed sets is reached.

5   Reconfiguration Model

After a failure or recovery, the system may change its configuration using the following types of *reconfiguration actions*:

- *changeReps(rsi, C, c)*: for replicated software instance *rsi*, change *rsi.computers* to set *C* of computers and, if *rsi* uses a passive replication protocol, change *rsi.primary* to computer *c*.
- *stop(si)*: stop running software instance *si*
- *stopRep(rsi)*: stop running replicated software instance *rsi*
- *start(si)*: create and start running software instance *si*
- *move(si, c)*: move a migratable software instance *si* to computer *c*, by migrating its volatile and persistent state, redirecting its network connections, etc.

New instances of "stateful" software (i.e., software *sw* satisfying *sw.persisState* $\lor \neg$ *sw.resumable*) cannot be started during reconfiguration, except that new members can be added to existing replicated instances of stateful software, provided the persistent state, if any, is small (hence can be copied quickly to new replicas). This reflects the assumption that a software component's state must be maintained continuously from the beginning of execution. (If the state did not need to be maintained from the beginning, then there would be no harm in losing it due to a failure and starting a fresh instance of the software, contradicting the definition of stateful software.) Since replication is used only for stateful software, this implies reconfiguration does not create new replicated software instances. Relaxing these assumptions and adding a *startRep* action for use when appropriate is straightforward.

We define a *reconfiguration relation reconfig(cfg,cfg',fs)* that holds if configuration *cfg* with failed set *fs* can be reconfigured to configuration *cfg'*. The straightforward way to define this relation is to first define simpler transition relations describing the effect of each type of reconfiguration action, and then define this relation to hold if there exists a sequence of reconfiguration actions that leads from state $\langle cfg, fs \rangle$ to state $\langle cfg', fs \rangle$.

Unfortunately, this leads to inefficient implementations that search over many sequences of actions, likely including many that lead to the same configuration.

We overcome this difficulty by giving an alternate characterization of the reconfiguration relation as a logical formula that directly expresses the possible differences between *cfg* and *cfg'*; the sequence of actions that produces those differences is implicit. This characterization supports efficient implementation. Both definitions of the reconfiguration relation are given in detail in [12].

A state $\langle cfg, fs \rangle$ is *valid* if *cfg* is valid and does not contain dead software instances, i.e., $cfg = removeDead(cfg, fs)$. A state $\langle cfg, fs \rangle$ is an *initial state* if it is valid and *fs* is empty.

Reconfiguration enables achieving a desired level of resilience with fewer resources. This benefit arises in a variety of situations, including the following. (1) The software providing a functionality is not stateful and not replicated, and the computer running it can fail. Reconfiguration can start a new instance of the software on a different computer, or a new instance of a different software providing that functionality, e.g., if the other computer has a different CPU architecture or OS, or if fewer computational resources are available. (2) The software providing a functionality is passively replicated and one or more replicas can fail. Reconfiguration is needed to replace the failed replicas with new ones, to tolerate subsequent failures. (3) A majority of replicas providing an actively replicated service can fail but not simultaneously. Reconfiguration can add new replicas after each set of simultaneous failures, to ensure a reconfiguration quorum is live when a subsequent failure occurs.

6   Resilience and Configuration Problems

This section defines resilience, along with auxiliary concepts, and then defines the core configuration problems.

A *resilience requirement* consists of a failure model *fm* and a set *critFns* of *critical functionalities*, i.e., functionalities that must be continuously available in every scenario involving failures consistent with *fm*.

A failed set *fs* is *maximal* with respect to a failure model *fm* if no strict superset of *fs* is consistent with *fm*; in other words, while all components in *fs* are failed with the specified failure types, no additional failures will happen.

Given a system model and a resilience requirement with failure model *fm* and critical functionalities *critFns*, we say that a state with configuration *cfg* and failed set *fs* is *resilient*, denoted *resilient*(*cfg*, *fs*, *fm*, *critFns*), if (1) all critical functionality is available in that state and, for each set of additional failures Δ consistent with *fm*, the system can be reconfigured after those failures to a new configuration *cfg'* such that (2a) all critical functionality is available in the resulting state $\langle fs \cup \Delta, cfg' \rangle$, and (2b) the resulting state is resilient. The base cases of this recursive definition of resilience correspond to maximal failed sets with respect to *fm*.

Note that our definition of resilience does not explicitly consider recoveries. Considering recoveries is unnecessary, because they are not guaranteed to occur and therefore cannot be relied on to provide resilience.

Theorem 1. *Considering only worst-case failed sets in the definition of resilience—i.e., replacing nextFS with nextFSworst—leads to an equivalent definition of resilience.*

Informally, this holds because a system that is resilient to a set of failures occurring simultaneously is also resilient to the same failures occurring in sequence (e.g., one at a time). The proof is straightforward.

The first core configuration problem is a decision problem. Given a system model, a resilience requirement with failure model *fm* and critical functionalities *critFns*, and a failed set *fs*, the *resilient configuration problem* is to find a configuration *cfg* such that the state $\langle cfg, fs \rangle$ is resilient with respect to *fm* and *critFns* if one exists, otherwise report "not resilient". Typically we want to find a resilient initial state, i.e., typically $fs = \emptyset$.

Some resilient configurations may be preferable to others, for example, because they require less hardware, use less power, or provide higher QoS. We capture the preference in a *quality metric* on configurations, which is a function from configurations to a totally ordered set. A sample quality metric is $quality(cfg) = \langle QoS(cfg), cost(cfg) \rangle$, where $QoS(cfg)$ is the number of instances of preferred software components (i.e., *sw.preferred* = true) running in *cfg*, and *cost(cfg)* is the total number of instances of software components running in it, counting a replicated software instance *rsi* as $|rsi.computers|$ instances. We use lexicographic ordering on these tuples.

To take the preference into account, we extend the decision problem to an optimization problem. Given a quality metric on configurations, the *best resilient configuration problem* is the same as the resilient configuration problem, except the goal is to find a configuration that has the highest quality among all resilient configurations.

7   Algorithm

A straightforward implementation of our framework that performs a naive search for resilient configurations or best resilient configurations is about 140 lines of DistAlgo code, ignoring imports, type declarations, and trivial code for initialization, logging, and printing. Sections 7.1 and 7.2 below describe two significant optimizations to the algorithm. We also implemented an optimized algorithm for *nextFSworst*, which avoids generating and then dropping non-worst-case failed sets in some cases. These optimizations add about 190 lines, ignoring trivial code for logging, etc.

7.1   Optimization: Avoid Irrelevant Configurations

Our algorithm starts by generating two sets of configurations: a set *initCfg* of initial configurations to use for the top-level search for a resilient one, and a set *allCfg* of valid configurations to use repeatedly when searching for the new configuration *cfg′* in case (2b) of the definition of resilience.

Our algorithm for generating *allCfg* is optimized to avoid enumerating invalid configurations as well as valid configurations that are *irrelevant* in the sense that they do not affect the answers to the decision problem, and they do not affect the answers to the optimization problem provided the configuration quality metric does not prefer expensive (high-resource) configurations. A configuration is irrelevant if it lacks software instances that provide critical functionalities, or contains too few replicas in replicated software instances (relative to the number of possible simultaneous failures and the quorum sizes of the replication algorithm), so that it clearly does not satisfy the conditions on *cfg′* in the definition of resilience, or if it has too many software instances, or too many replicas in replicated software instances, so that it clearly uses more resources than other configurations satisfying the conditions on *cfg′*. A detailed definition of irrelevant configuration is in [12].

Our algorithm for generating *initCfg* uses a broader definition of irrelevant configurations that extends the above definition with the following clauses. (1) For replicated software instance *rsi* with a passive replication protocol, if all computers in *rsi.computers* are *equivalent* (i.e., have the same values for all attributes listed in Sect. 3), then one is chosen (arbitrarily) as the primary, and configurations that differ only by having a different primary for *rsi* are irrelevant. (2) A configuration *cfg* is irrelevant if it contains a replicated software instance *rsi* whose replica set *rsi.computers* contains a computer that cannot run *rsi.sw* in configuration *cfg* due to unsatisfied dependence, compatibility, capacity, or distribution constraints.

7.2 Optimization: RS-Equivalence Quotient Reduction

A quotient reduction [2] is a technique for reducing the number of states of a system that need to be considered during verification, by exploiting an equivalence relation on states. The equivalence relation must be designed to preserve the system behavior and the properties being verified. This ensures that it is sufficient to consider only one representative of each equivalence class when verifying those properties. Symmetry reduction is a well-known type of quotient reduction, where the equivalence relation is based on a symmetry relation; clause (1) in the extension of "irrelevant" used when generating *initCfg* is a simple example of a symmetry reduction.

We developed a quotient reduction based on a novel equivalence relation called *relocatable-software equivalence* (RS-equivalence). A software component is relocatable (defined formally below) if it can be relocated during reconfiguration by starting a new instance on a different computer. The main idea is that two configurations are relocatable-software equivalent(RS-equivalent) if they differ only in the computers on which relocatable software components are running— in other words, if one of those configurations can be obtained from the other by relocating instances of relocatable software components. Intuitively, if a relocatable software component can be run on either of two computers, then we can arbitrarily choose to run it on one of them, and ignore the possibility of running it on the other, because if the chosen computer fails, we can start a new instance of that software component on the other computer during reconfiguration.

RS-equivalence applies to systems where the dependency relation between functionalities is acyclic, with the dependency relation defined by: functionality fn_1 depends on functionality fn_2 if some software component that provides fn_1 requires fn_2. This condition is typically satisfied in practice, since software systems typically do not have recursive (cyclic) dependencies between services. Generalizing RS-equivalence to allow cyclic dependencies is a direction for future work.

A software component *sw* is *relocatable* in configuration *cfg*, denoted *reloc(sw, cfg)* if: (1) *sw* is resumable, does not have persistent state, and is not slow-starting; (2) if any software component in *cfg* depends on *sw*'s functionality, then *sw* supports remote use; and (3) software components in *cfg* on which *sw* depends support remote use. The first condition ensures that, during reconfiguration, the software can be stopped and then re-started on a different computer. The second and third conditions ensure that software dependencies cannot prevent relocation of this software during reconfiguration.

Configurations cfg_1 and cfg_2 are *RS-equivalent* if they are valid and, letting $R_i = \{si \in cfg_i.SI \mid reloc(si.sw, cfg_i)\}$, there exists a bijection f from R_1 to R_2 such that:

$\forall si_1 \in R_1 :$
$si_1.sw = f(si_1).sw \wedge si_1.sw.devices \cap si_1.computer.devices = si_1.sw.devices \cap f(si_1).computer.devices$
$\wedge \quad cfg_1.SI \setminus R_1 = cfg_2.SI \setminus R_2$ (The configurations contain the same instances of non-relocatable software components.)
$\wedge \quad cfg_1.RSI = cfg_2.RSI$

Note that the computers on which a relocatable software component sw is running in cfg_1 and cfg_2 do not need to be identical. They do need to have the same types of integrated devices used by sw; otherwise failure of a stand-alone device could break the equivalence, i.e., cause sw to be runnable on one of those computers but not the other.

Lemma 1. *If configurations cfg_1 and cfg_2 are RS-equivalent, then for all failure states fs such that $\langle cfg_1, fs \rangle$ and $\langle cfg_2, fs \rangle$ are valid, the same functionality is available in these states.*

Proof Sketch. The proof is by induction on the set of available functionalities in each configuration, partially ordered by the dependency relation between functionalities. The proof is straightforward. □

Lemma 2. *RS-equivalence is an equivalence relation.*

Proof Sketch. The proof is straightforward.

□

Theorem 2. *RS-equivalence preserves resilience. Specifically, for every failure model fm, set of critical functionalities critFns, failed set fs consistent with fm, and valid configurations cfg_1 and cfg_2, if states $\langle cfg_1, fs \rangle$ and $\langle cfg_2, fs \rangle$ are valid, and cfg_1 and cfg_2 are RS-equivalent, then resilient(cfg_1,fs,fm,critFns) iff resilient(cfg_2,fs,fm,critFns).*

Proof Sketch. We show that resilience of cfg_1 implies resilience of cfg_2; the opposite implication follows from the same reasoning, due to the symmetry in the statement of the theorem. In short, we need to show that (1) all functionality in *critFns* is available in cfg_2, and (2) for each failure state fs' in $nextFS(fm, fs)$, there exists a resilient configuration cfg' to which $removeDead(cfg_2, fs')$ can reconfigure. Conclusion (1) follows from the assumption that cfg_1 is resilient and Lemma 1. Consider conclusion (2). Resilience of cfg_1 implies that, for each failure state fs' in $nextFS(fm, fs)$, there exists a resilient configuration cfg' to which $removeDead(cfg_1, fs')$ can reconfigure, using some sequence S of reconfiguration actions. It is reasonably straightforward to show, using the premises of the theorem and the definition of RS-equivalence, that $removeDead(cfg_2, fs')$

can reconfigure to $removeDead(cfg_1,fs')$, by using stop and start actions to relocate relocatable software instances that are on different computers in $removeDead(cfg_1,fs')$ and $removeDead(cfg_2,fs')$, and then it can reconfigure to cfg' using actions S.   □

7.3   Optimized Search-Based Algorithm

Our search-based algorithm for finding resilient configurations, incorporating the above optimizations, works as follows:

1. Generate the sets $initCfg$ and $allCfg$ of all relevant initial configurations and all relevant configurations, respectively.
2. Compute equivalence classes with respect to RS-equivalence for $allCfg$. Specifically, create a dictionary D that maps each configuration cfg to a configuration that uniquely represents cfg's equivalence class.
3. Create a set $initCfg'$ containing the representatives of the equivalence classes of configurations in $initCfg$.
4. For each configuration cfg in $initCfg'$, check $resilient(cfg, \emptyset, fm, critFns)$.

To evaluate $resilient(cfg, fs, fm, critFns)$, first check that all critical functionalities are available in state $\langle cfg, fs \rangle$, and then for each failed set fs' in $nextFSworst(fm, fs)$:

1. Compute the set $next$ of next configurations by iterating over $allCfg$ and adding each configuration cfg_1 such that $reconfig(removeDead(cfg,fs'), cfg_1, fs')$ and $next$ does not already contain a member of cfg_1's equivalence class.
2. Check whether $resilient(cfg_1, fs', fm, critFns)$ holds for some cfg_1 in $next$. If not, return false.

If the loop exits normally (without returning false), return true. The recursion terminates when $nextFSworst$ returns an empty set. The algorithm can easily be extended to compute and store the sequence of reconfiguration actions needed to transition from a state to a resilient successor state; this provides the reconfiguration policy.

To find resilient states that optimize a given configuration quality metric, we sort $initCfg'$ in descending order by configuration quality, and check resilience of the configurations in that order. With the extension to store reconfiguration actions, in the algorithm for evaluating $resilient$, among the configurations cfg_1 found to be resilient in step 2, we store the sequence of reconfiguration actions that leads to a configuration of highest quality configuration among those. A straightforward extension is to add a notion of reconfiguration cost (e.g., based on reconfiguration time) and store the lowest-cost sequence of reconfiguration actions from cfg to cfg_1.

8   Evaluation

We evaluated our approach on a model of an autonomous driving system. The software architecture models Autoware, a prominent open-source software stack for self-driving vehicles. The computing platform models QualComm's Snapdragon Ride platform, which uses multiple smaller embedded computers with ARM processors, for greater scalability and power-efficiency than a single large computer.

The hardware model includes several sensors (camera, GPS, IMU, LIDAR, and radar) and two or more quad-core ARM-based embedded computers running Safe RT Linux, a real-time OS. The embedded computers have enough RAM that it is not a limiting factor. We assume the network is also real-time, hence the system is synchronous.

We modeled five main software components with the following functionalities: (1) perception: performs object detection, tracking, and prediction using data from radar, LIDAR, and camera; (2) localization: estimates the location of the vehicle using data from GPS, IMU, camera, etc.; (3) planning: produces a trajectory towards a provided goal, starting from the current position and velocity; (4) control: generates commands (for steering, braking, etc.) that will make the vehicle follow a given trajectory; (5) vehicle interface: interfaces the autonomous driving software with a specific vehicle model.

The system includes preferred Safe RT Linux versions of all five software components. Each software component requires one core, except the perception component, which requires two cores. Perception maintains critical volatile state, which is needed for object tracking and prediction, and hence this software component is not resumable; the other four software components are resumable. None of the software components is slow-starting, and none maintain critical persistent state. The primary-backup protocol is used for replication, since the system is synchronous, hardware resources are limited, and primary-backup requires fewer replicas to tolerate a given number of failures. We consider one type of failure: crashes of embedded computers.

We applied our optimized search-based algorithm to two versions of this system with different sets of hardware and software components, described below. We report the time needed to find *all* resilient initial configurations—or, when RS-equivalence reduction is used, to find all equivalence classes of resilient initial configurations—and then sort them by quality, using the sample configuration quality metric in Sect. 6. This comprehensive result is not required by the problem definitions in Sect. 6 but forces comprehensive exploration of the search space, and eliminates potential variation in running time due to non-determinism in iteration orders, which can cause the first resilient configuration to appear earlier or later in the search. All experiments were run on a Mac-Book Air M1 (2020) with 8 GB RAM running macOS 11.6, Python 3.7.12 and DistAlgo 1.1.0b15, with garbage collection disabled for more consistent running times. We validated consistency of running times by repeating several experiments three times and noting that all measured running times were within 5% of the mean.

8.1   Example 1: Failover to Laptop

In this example, the hardware model is extended with a laptop running Linux. It is not normally part of the autonomous driving system, but can be used for some autonomous driving functionality in case of failures. The software model is extended with non-preferred Linux versions of the planning and control components. Localization and perception are less suitable for migration to the laptop, because they process streams of sensor data and hence require a high-bandwidth connection to the vehicle's network. The vehicle interface component is by nature hardware-specific and requires direct connection to the vehicle's network. All five functionalities are critical.

We illustrate the results of the resilience analysis by describing one highest-quality resilient initial configuration in a system with two embedded computers, $c0$ and $c1$, and how it reconfigures after a failure. In the initial configuration, unreplicated instances of the control and vehicle interface software run on $c0$, unreplicated instances of the planning and localization software run on $c1$, and a replicated instance of the perception software runs on $c0$ (the primary) and $c1$.

Upon failure of $c0$, the system reconfigures by stopping the instance of planning software on $c1$, starting an instance of the vehicle interface software on $c1$, starting the Linux versions of the control and planning software on the laptop, making $c1$ the primary replica for the perception software, and removing $c0$ as a replica for the perception software (there is only one replica at that point, but by continuing to represent it as a replicated software instance, we can easily add a replica if $c0$ recovers).

Table 1 presents the numbers of configurations and the numbers of equivalence classes of configurations generated. The first column reports N_C (the number of embedded computers) and N_F (the maximum number of failures of embedded computers, which we always take to be $N_C - 1$). The second and third columns report the sizes of *allCfg* and *initCfg*, respectively. The remaining columns report the numbers of equivalence classes of configurations in *allCfg*, *initCfg*, and *resilCfg*, respectively, where *resilCfg* is the set of resilient initial configurations found.

Table 2 presents the running times of three versions of the algorithm, with: no quotient reduction, abbreviated "no Q.R."; partial RS-equivalence quotient reduction (i.e., for *initCfg* only, in step 3 of the pseudocode for search in Sect. 7.3), abbreviated "partial Q.R."; and full RS-equivalence quotient reduction (for *initCfg* and *next*), abbreviated "full Q.R.". The running time is partitioned into generate time (the time taken to enumerate all relevant configurations, and to generate equivalence classes based on RS-equivalence if applicable) and analyze time (the rest of the running time). "T-O" (time-out) indicates the experiment was aborted after 1 h.

We see that partial Q.R. drastically reduces the running time for systems with more computers, e.g., by a factor of 30 for Example 2 with three computers, and even larger factors with more computers (though the exact speedup ratios in these cases were not measured due to the 1-hour time-out). Even with this optimization, the running time grows quickly with the number of computers,

158   S. D. Stoller et al.

indicating that there is opportunity for symmetry reductions beyond the simple one we are using in the definition of irrelevant for *initCfg*.

We also observe that the algorithm with full Q.R. has comparable running time as the algorithm with partial Q.R. The full reduction further reduces the number of explored configurations, and therefore is still beneficial if RAM is at a premium. However, the time saved by exploring fewer states is balanced by the overhead of applying the reduction to *next* at every step.

Table 1. Numbers of configurations analyzed for Examples 1 (left) and 2 (right).

| $\langle N_C,$ | # of configs | | # of equiv. classes | | |
$N_F \rangle$	*allCfg*	*initCfg*	*allCfg*	*initCfg*	*resilCfg*
$\langle 2, 1 \rangle$	256	204	36	27	9
$\langle 3, 2 \rangle$	4,332	2,607	108	63	9
$\langle 4, 3 \rangle$	36,992	17,648	288	135	9
$\langle 5, 4 \rangle$	224,720	88,015	720	279	9

| $\langle N_C,$ | # of configs | | # of equiv. classes | | |
$N_F \rangle$	*allCfg*	*initCfg*	*allCfg*	*initCfg*	*resilCfg*
$\langle 2, 1 \rangle$	162	128	24	18	6
$\langle 3, 2 \rangle$	2,520	1,512	72	42	6
$\langle 4, 3 \rangle$	20,720	9,872	192	90	6
$\langle 5, 4 \rangle$	123,120	48,190	480	186	6

Table 2. Running times, in seconds, for Examples 1 (left) and 2 (right). "gen." abbreviates "generate".

| $\langle N_C,$ | no Q.R. | | partial Q.R. | | full Q.R. | |
$N_F \rangle$	gen.	analyze	gen.	analyze	gen.	analyze
$\langle 2, 1 \rangle$	0.1	2.0	0.1	0.3	0.1	0.1
$\langle 3, 2 \rangle$	1.4	171.1	1.4	9.8	1.6	6.1
$\langle 4, 3 \rangle$	10.8	T-O	11.6	110.2	13.3	106.7
$\langle 5, 4 \rangle$	69.7	T-O	67.3	1,263.2	84.5	1,283.4

| $\langle N_C,$ | no Q.R. | | partial Q.R. | | full Q.R. | |
$N_F \rangle$	gen.	analyze	gen.	analyze	gen.	analyze
$\langle 2, 1 \rangle$	0.1	0.6	0.1	0.1	0.1	0.1
$\langle 3, 2 \rangle$	0.5	74.4	0.6	2.0	0.6	1.9
$\langle 4, 3 \rangle$	4.5	T-O	4.51	29.4	5	29.2
$\langle 5, 4 \rangle$	24.5	T-O	26	367.8	31.8	347.4

8.2   Example 2: Failover to Smartphone-Guided Human Operator

In this variation, there is a human operator who can take over some driving functions in case an embedded computer fails. The hardware and software components are mostly the same as in the previous example, except without the Linux laptop and Linux versions of software components, and with a smartphone and a non-preferred manual-control software component that runs on Android smartphones. The manual-control software uses the smartphone's display and speaker to convey driving instructions to follow the path computed by the planning component, in a similar way as Google Maps. When the manual-control software is used for control, the vehicle interface component is not needed; therefore, vehicle interface functionality is not critical in this example. The other four functionalities are critical.

After a failure, the system reconfigures in a similar way as Example 1, by starting a less-preferred instance of the control software on the smartphone, starting any needed instances of other software components on the surviving embedded computer, etc.

Performance results for Example 2 appear in Tables 1 and 2. The above performance observations for Example 1 apply to Example 2, too. We also note that resilience analysis for Example 2 explores fewer configurations and is faster than for Example 1, because Example 2 involves only one software component with a less-preferred alternative version, while Example 1 involves two.

9 Conclusion

We developed a general automated framework for making systems resilient by optimized adaptive distribution and replication of software components, including a model-finding algorithm that finds resilient states and reconfiguration policies using state-space exploration with multiple optimizations including a quotient reduction based on a novel equivalence relation.

Directions for future work include: modeling and analysis of network failures and partitions; probabilistic failure models and probabilistic resilience properties; more detailed timing analysis to consider response-time requirements, failover latency, etc.; partially-ordered quality metrics that capture trade-offs between cost and multiple aspects of resilience (e.g., resilience to data corruption due to transient failures as well as resilience to crashes); and new performance optimizations, including symmetry reductions that reduce the set of failed sets that are considered. We also plan to explore new case studies, especially involving multi-agent systems such as collections of robots or drones, with functionalities replicated across agents and possibly base stations.

References

1. Daidone, A., Renier, T., Bondavalli, A., Schwefel, H.P.: Optimal configuration of fault-tolerance parameters for distributed replicated server access. Int. J. Critical Comput.-Based Syst. **4**(2), 144–172 (2013). https://doi.org/10.1504/IJCCBS.2013.056493
2. Emerson, E.A.: Model checking: Theory into practice. In: Foundations of Software Technology and Theoretical Computer Science (FST TCS 2000), pp. 1–10. Springer (2000)
3. Lamport, L.: The part-time parliament. ACM Trans. Comput. Syst. **16**(2), 133–169 (1998)
4. Lamport, L., Malkhi, D., Zhou, L.: Reconfiguring a state machine. SIGACT News **41**(1), 63–73 (2010). https://doi.org/10.1145/1753171.1753191
5. Li, X., Houshmand, F., Lesani, M.: HAMRAZ: resilient partitioning and replication. In: 2022 IEEE Symposium on Security and Privacy (SP), pp. 2267–2284 (2022). https://doi.org/10.1109/SP46214.2022.9833661
6. Liu, Y.A., Stoller, S.D., Lin, B.: From clarity to efficiency for distributed algorithms. ACM Trans. Programm. Lang. Syst. **39**(3), 12:1–12:41 (May 2017)

7. Machida, F., Kawato, M., Maeno, Y.: Redundant virtual machine placement for fault-tolerant consolidated server clusters. In: 2010 IEEE Network Operations and Management Symposium - NOMS 2010, pp. 32–39 (2010). https://doi.org/10.1109/NOMS.2010.5488431

8. de Moura, L., Bjørner, N.: Z3: An efficient SMT solver. In: Ramakrishnan, C.R., Rehof, J. (eds.) Tools and Algorithms for the Construction and Analysis of Systems, pp. 337–340. Springer, Heidelberg (2008)

9. Ongaro, D., Ousterhout, J.: In search of an understandable consensus algorithm. In: 2014 USENIX Annual Technical Conference, pp. 305–319. USENIX Association (2014)

10. Schneider, C., Barker, A., Dobson, S.: A survey of self-healing systems frameworks. Softw. Practice Exp. **45**(10), 1375–1398 (2015). https://doi.org/10.1002/spe.2250

11. Stoicescu, M., Fabre, J.C., Roy, M.: Architecting resilient computing systems: a component-based approach for adaptive fault tolerance. J. Syst. Architect. **73**, 6–16 (2017). https://doi.org/10.1016/j.sysarc.2016.12.005

12. Stoller, S.D., Jayasankar, B., Liu, Y.A.: Resilience through automated adaptive configuration for distribution and replication (2025). https://doi.org/10.48550/arXiv.2506.10248

13. Weyns, D.: Introduction to Self-Adaptive Systems: A Contemporary Software Engineering Perspective. Wiley (2020)

14. Wong, T., Wagner, M., Treude, C.: Self-adaptive systems: a systematic literature review across categories and domains. Inf. Softw. Technol. **148**, 106934 (2022). https://doi.org/10.1016/j.infsof.2022.106934

On-the-Fly Cone-of-Influence Reduction for Model Checking Concurrent Software

Csanád Telbisz[✉][iD], Levente Bajczi[iD], Dániel Szekeres[iD], and András Vörös[iD]

Department of Artificial Intelligence and Systems Engineering, Budapest University of Technology and Economics, Budapest, Hungary
csanadtelbisz@edu.bme.hu, {bajczi,szekeres,vori}@mit.bme.hu
https://ftsrg.mit.bme.hu

Abstract. Calculating successor states in SMT-based software model checking is a costly task that often requires solving an SMT problem. However, in many cases, the evaluation of a program statement has no effect with respect to the verified property. Successor state calculation can be simplified in such cases. Several algorithms exist such as the cone-of-influence reduction that statically analyze the model and eliminate irrelevant variables from the model. In concurrent software, however, it is common for the result of a statement to be used in one interleaving of threads while unused in another. Algorithms that statically analyze the model cannot simplify such statements. Our on-the-fly approach detects whether a statement can be simplified during the state space exploration based on the current state of each process. Evaluation results show that our algorithm can simplify around 20% of all statements on average over a large set of benchmark programs while reducing the time of successor state calculation by more than 30% on average.

Keywords: cone-of-influence · concurrency · abstraction · data-flow

1 Introduction

Model checking has been a field of active research in recent decades as it is one of the most powerful software verification techniques. Model checking algorithms face the state space explosion problem: the state space of software systems grows exponentially with the number of variables [24]. Concurrency further increases the complexity due to the great number of possible thread interleavings. Various techniques have been developed to tackle this vast complexity. Partial order reduction avoids exploring parts of the state space when it is guaranteed that an equivalent thread interleaving is explored for each avoided trace [44]. Abstraction reduces the size of the state space by ignoring some details of the original problem [22,23]. Counterexample-guided abstraction refinement (CEGAR) iteratively refines the abstraction until the desired property can be verified [22]. Other abstraction-based techniques like the cone-of-influence (COI) reduction or program slicing eliminate model elements irrelevant to the verified property [10,33].

G. Ernst and K. Y. Rozier (Eds.): SPIN 2025, LNCS 15945, pp. 161–181, 2026.
https://doi.org/10.1007/978-3-032-06847-7_9

Existing cone-of-influence and slicing techniques choose eliminable variables or statements using static data-flow analysis based on the control-flow of the program [10,33]. In concurrent programs, this kind of elimination is often ineffective due to the communication between threads and the many possible thread interleavings. To address this, we propose an algorithm that can eliminate statements based on the current local states of concurrent threads. Whereas the COI reduction simplifies the model by eliminating completely redundant variables (redundant in all thread contexts) regarding the verified property [10], our approach identifies and simplifies statements on-the-fly that are redundant in the current state of concurrent threads with respect to the verified property. Thus, our method is more fine-grained: it can eliminate statements in certain contexts even if the statement cannot always be ignored. This is particularly useful when a statement is relevant in one interleaving of threads while it is redundant in another: we eliminate it in cases when it is redundant. As our algorithm takes its main advantage from the local states and interleaving of threads, we focus our presentation on concurrent programs. While our method could also be used for sequential programs, it loses its advantage over other techniques in that case.

As an example, take the program with two threads from Fig. 1a. Let us take a state s from the state space of the program where process p_2 has executed the statement y := x previously (i.e., this statement can be found on the path from the initial state to s). Observe that the value of x cannot be read by any statement of any thread reachable from s in the state space. Thus, it is unnecessary to evaluate x := 1 or x := 0 after s. Our algorithm detects such situations and eliminates such statements. Note that a traditional COI algorithm could not eliminate the statement x := 1 as its result may be used.

Our statement simplification method is motivated by the considerable runtime of calculating successor states in SMT-based state space exploration [12,32]. We strive to dynamically identify as many redundant statements as possible in the current exploration context. For this, we build a data-flow graph and update it based on the current thread interleaving during the state space exploration to reflect the individual states of each process. Before evaluating a statement (i.e., calculating the successor of the current state with respect to this statement), we check using the data-flow graph whether any other statement can use the result of the statement. We target reachability properties; thus, we are interested in whether the result of the statement is used transitively by a conditional statement, as only conditionals can directly influence whether some marked error locations in the model are reachable. This can be decided by a traversal of the data-flow graph. Redundant statements are eliminated, sparing the time of successor state calculation in such cases. We formulate our algorithm for abstract state space exploration and exploit information about the current abstraction to reduce the number of edges in the data-flow graph.

To further motivate our approach, it is possible to achieve exponential gains in the number of evaluated statements by using our novel algorithm. Consider the example from Fig. 1b with $2N+1$ processes. The safety of the program can be proven with abstraction by only tracking the predicates $z \bmod 2 = 0$ and $x = 0$

Process p_1	Process p_2
`x := 1`	`y := x`
`y := 1`	`x := 0`
`assert(y=1)`	

(a) Simple example

initially: `x:=y:=z:=0`

Process p_0	Processes $p_1 - p_N$
`repeat N times:`	`y := y+1`
`  z := z+2*y`	
`if z mod 2 = 0:`	
`  x := 0`	Processes $p_{N+1} - p_{2N}$
`else:`	
`  x := 1`	`y := y*y`

finally: `assert(x*y=0)`

(b) Example with possible exponential gain

Fig. 1. Motivational examples for demonstration

about our variables: $z \bmod 2 = 0$ is an invariant of the loop of p_0, so x gets 0 that satisfies the assertion. Processes $p_1 - p_{2N}$ have $2N!$ interleavings not considering the statements of the loop of p_0; together there are even more interleavings. This is indeed a difficult task for verifiers: we tried state-of-the-art tools (such as ULTIMATE [38] and DARTAGNAN [41]), and they cannot solve the problem for $N > 4$ within a reasonable time. However, our algorithm notices that we do not track any information about y, so the results of statements writing y are not used in this abstraction: so, our approach eliminates all of these statements ($2N! * 2N$ statements exactly). Our approach also enables existing partial order reduction algorithms [1,3,30] to reduce the number of explored interleavings exponentially that otherwise would have to explore all interleavings. Our algorithm achieves this by eliminating the source of dependency between statements.

Contributions. We take the base idea of the cone-of-influence reduction one step further by deciding on-the-fly during state space exploration whether the result of a statement can be used later. We present a novel algorithm for identifying redundant statements using an abstract dynamically updated data-flow graph. Furthermore, we discuss the necessary additions in an iterative abstraction-refinement verification scheme (namely, CEGAR). We have implemented and evaluated our algorithm in the abstraction-based model checking tool THETA [48].

2   Related Work

Several works aim to simplify the model by eliminating redundant model elements based on data-flow analysis [10,28,33,36,42,43]. These techniques only statically analyze and simplify the input model which is limited compared to our on-the-fly data-flow analysis. These static approaches have the advantage that they have to be executed only once before the state space exploration while our algorithm is performed at each successor state calculation. On the other hand, our experiments in Sect. 5 show that our approach does not have a significant overhead, so it is worth running our algorithm several times (i.e., once for each state transition during the state space exploration) to eliminate further statements that static techniques cannot eliminate.

There are dynamic program slicing techniques as well [33,39]. However, *dynamic*, in those contexts, means that these techniques use actual input values or already discovered error traces for slicing [5]. These techniques do not take advantage of the local states and interleaving of threads (most of them formulated for sequential programs [39]) which is the basis of our approach.

Many algorithms have been developed for model checking concurrent programs that reduce the number of explored thread interleavings such as partial order reduction or maximum causality reduction [1,4,37]. Some of these techniques leverage abstraction-related information to achieve further reduction [8,29,30], but they do not use a complex data-flow analysis to further reduce the number of dependent program statements. Some other works perform dynamic data-flow analysis in various ways to improve the reduction potential of these techniques [4,6,20,37,45], though they only use data-flow analysis to reduce the number of explored interleavings and not to simplify statements. These techniques take explored traces and discover redundant statements within these traces: they use this information to explore even less interleavings (e.g., by ignoring these statements when calculating a dependency relation [6]). The works [4,37] discover causality connections and build causality constraints between statements (events) of a trace and simplify these formulae by eliminating irrelevant statements which is a similar concept to our approach. However, they only use this idea to simplify these constraints, but they still completely explore traces first. So our approach could achieve further reduction in these cases as well. In other words, these works aim to reduce the size of the explored state space whereas our purpose is to accelerate the exploration of a (reduced) state space by skipping the evaluation of certain program statements. Our algorithm is orthogonal to these techniques and could be applied on top of them to further improve the performance by eliminating further model elements.

The combination of abstraction-based and slicing techniques have already been investigated [18,27]. Those approaches enhance the refinement step of the counterexample-guided abstraction refinement approach by slicing infeasible error traces found during state space exploration. On the other hand, our technique works in the state space exploration phase of CEGAR. Therefore, the two techniques are orthogonal and have a different purpose. We include some remarks about the consistent use of our technique with refinement methods in Sect. 4.3.

Dynamic pruning and a so-called *dynamic cone-of-influence* algorithm is also introduced in [21]. However, it is just a coincidental name collision: they use it in the context of fault tree analysis to lazily construct fault trees and compute minimal cut sets. It has nothing to do with concurrent programs.

3   Preliminaries

This section introduces the basic concepts and notions regarding abstract state space exploration necessary for the presentation of the proposed technique.

3.1   Computation Model

In this paper, we assume a computation model of concurrent programs where processes (threads) communicate via shared variables. We assume a sequential consistency memory model. Though it would be easy to incorporate extra features into the model (such as heap memory, dynamic thread creation or termination, and synchronization primitives), we strive to keep our presentation simple and skip these details. Our implementation for the evaluation naturally supports these features. We represent concurrent programs by control-flow automata (CFA) [15]: each process has its own (conventional) CFA representation.

Definition 1. *A multi-threaded CFA is a tuple (V, P), where*

- *V is a set of (global) variables,*
- *P is a set of processes. A process is a tuple $p = (L, l_0, A, E)$, where:*
  - *L is a set of control locations with $l_0 \in L$ as the initial location,*
  - *A is a set of statements,*
  - *$E \subseteq L \times A \times L$ is a set of transitions. A transition is a directed edge with a source control location, a target control location, and one statement.*

Each variable $v \in V$ has a domain D_v (the possible values for v), and possibly an initial value from its domain. A statement can be a deterministic assignment $(v = expr)$, a non-deterministic assignment (*havoc v*) where the new value of v can be anything from its domain, or a guard condition ($[cond]$). For the verification of reachability properties, some CFA locations are marked as error locations: the program is safe if no error location can be reached by any of its processes. We define transition systems (state spaces) as follows:

Definition 2. *A transition system is a tuple (S, A, T, I), where S is a set of states, A is a set of actions, $T \subseteq S \times A \times S$ is a set of transitions, and $I \subseteq S$ is a non-empty set of initial states.*

In this paper, actions correspond to statements as introduced above. An action α is an *outgoing* action from a state s if there is a transition $(s, \alpha, s') \in T$ for some $s' \in S$. We use the following notations:

- $\alpha(s) = \{s' \in S : \quad \exists (s, \alpha, s') \in T\}$,
- *outgoing(s)* denotes the set of outgoing actions from s,
- *vars(α)* denotes the set of variables referenced by α,
- *written(α)* and *read(α)* is the set of variables written/read by α, respectively.

The state space of a program is a transition system where a state stores the CFA locations of all processes and the values of all variables. An action of a transition corresponds to a statement of a single process (processes step asynchronously). We use the Greek alphabet for actions, and we write p_α for the process of action α. Note that *written(α)* has a single item for deterministic and non-deterministic assignments, and it is an empty set for a guard condition. We denote the control location of process p in state s by $s(p)$, and the value of variable v in state s by $s(v)$. We define an expression function for a state s based on the values of variables in s: $expr(s) := \bigwedge_{v \in V}(v = s(v))$. By $w = t_1...t_k$, we denote a transition sequence (or trace), and we use the following for the concatenation of transition sequences or transitions: $w.v$. We also refer to action sequences as traces. A state is an error state if any of the processes is in an error location in the state. A state s' is said to be reachable from a state s if there is a trace starting from s and ending in s'. If we have a trace from a state that leads to an error state, we call this trace an error trace.

3.2   Abstraction

An abstraction can be defined through an abstract domain, a precision, and a transfer function [14].

Definition 3. *An abstract domain is a tuple $(S, expr)$, where S is a set of abstract states[1], and $expr : S \mapsto FOL$ is an expression function mapping an abstract state to a first-order logic formula describing the state.*

We assume that CFA locations of all processes are explicitly tracked in all abstract domains (and refer to the location of process p in a state s by $s(p)$ as introduced earlier). An abstract state s represents a concrete state c denoted by $c \models s$ if $c(p) = s(p)$ for each process p, and $expr(c)$ implies $expr(s)$. In our notation, we use s for abstract states and c for concrete states. An abstract state is an error state if any process is in an error location.

An abstract trace $w = \alpha_1...\alpha_k$ from the abstract state s_0 ($s_0 \xrightarrow{\alpha_1} ... \xrightarrow{\alpha_k} s_k$) is *feasible* if w is also a trace in the concrete state space ($c_0 \xrightarrow{\alpha_1} ... \xrightarrow{\alpha_k} c_k$) with $c_i \models s_i$; otherwise, w is spurious from s_0. The abstract state space over-approximates the behavior of the concrete state space: if there is a trace w from a concrete state c, then w is also a trace in the abstract state space from all abstract states s with $c \models s$ [14].

The precision (Π) describes which aspects the abstraction keeps, defined differently for each abstract domain. The *variables of a precision $vars(\Pi)$* are the variables that may appear in abstract state expression formulae. As a consequence, the abstraction tracks no information about variables in $V \setminus vars(\Pi)$. The transfer function calculates the successor states of an abstract state with respect to a statement and a precision.

Two frequently used abstract domains are explicit-value [17] and predicate abstraction [31]. In explicit-value abstraction, an abstract state is defined by the CFA locations of processes and an abstract variable assignment. The precision is the subset of variables $\Pi \subseteq V$ that are explicitly tracked in this abstraction, therefore $vars(\Pi) = \Pi$, here. Values of other variables are unknown in all abstract states. The expression function of an abstract state is defined similarly to concrete states in Sect. 3.1: variables whose values are unknown in a state are simply omitted. The result of the transfer function is based on the strongest

---

[1] Abstract states are usually defined as a semi-lattice with a partial order [14], but we do not need those details for this paper, so we simplify.

post-operator under abstract variable assignment [17]. In predicate abstraction, an abstract state is defined by the CFA locations of processes and a combination of first-order logic (FOL) predicates [31]. The precision is a set of FOL predicates (e.g., x > 0, y = z) that are tracked in the abstraction; $vars(\Pi)$ is the set of variables appearing in the tracked predicates. The expression function of an abstract state is the combination of FOL predicates that describes the state [31].

4   Statement Reduction During Dynamic Analysis

This section presents a method for simplifying the statement of an action before calculating the successors of the current state with respect to the action. Basically, when there is no possible interleaving of threads from the current state where the value of a written variable is read by any relevant statement regarding the verified property, we do not evaluate the expression writing the variable.

4.1   Data-Flow Graph with Precision

First, let us formalize the connection between actions of the program when one action uses the result of another action.

Definition 4. *Let α, β be actions, and Π be the precision of the abstraction. We say that β* observes *α with precision Π if $written(\alpha) \cap read(\beta) \cap vars(\Pi) \neq \emptyset$.*

An action α is transitively observed *by an action β in a trace $w = w_1...w_n$ if there is a sequence of indices $i_1, ..., i_m$ $(1 \leq i_1 < ... < i_m \leq n)$ such that w_{i_j} is observed by $w_{i_{j+1}}$ for each $1 \leq j < m$, and $w_{i_1} = \alpha$, $w_{i_m} = \beta$.*

Note that the sequence of indices in the definition does not necessarily contain adjacent indices (i.e., i_{j+1} is not necessarily $i_j + 1$): for example, in the trace $x = 1, z = 1, y = x$, the last action transitively observes the first with indices $i_1 = 1$ and $i_2 = 3$ in the definition. Each action α transitively observes itself as the trace consisting of the single action α fulfills the conditions of the definition. Also note that this is an over-approximation of possible data-flow between α and β, since it is possible that a variable is rewritten before it is observed (e.g., actions $x = 1$, $x = 2$, $y = x$ in this order).

We build an abstract data flow graph whose nodes are statements of the program and a directed edge represents an observation between the connected nodes, i.e., the target action observes the source of the edge. There are two types of edges: in-process (**Direct**) and inter-process (**Indirect**) observation.

Definition 5. *An abstract data-flow graph is a tuple $G = (A, D, I, \Pi)$ where:*

- *A is the set of actions of the program (the nodes of the data flow graph),*
- *$D \subseteq A \times A$ is the set of direct observation edges: $(\alpha, \beta) \in D$ if β observes α with Π, $p_\alpha = p_\beta$, and β is reachable from α in the CFA of their process,*

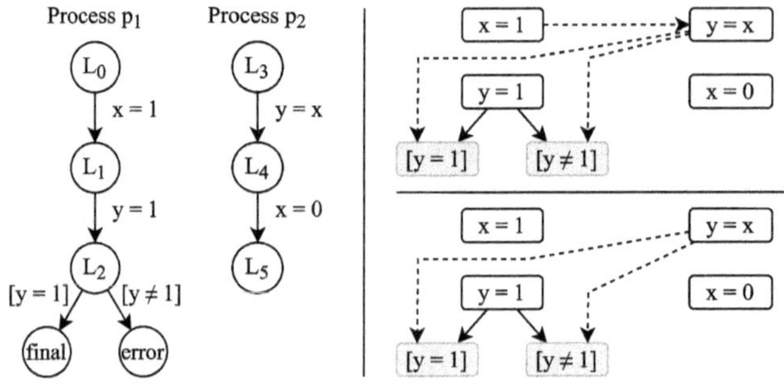

Fig. 2. CFA of two processes and data-flow graphs with different precisions

- $I \subseteq A \times A$ *is the set of indirect observation edges:* $(\alpha, \beta) \in I$ *if* β *observes* α *with* Π *and* $p_\alpha \neq p_\beta$.[2]

The data-flow graph can be precomputed for the state space exploration and the same data-flow graph can be used for the full state space exploration without the need for updating it as long as the precision of the abstraction is the same. To collect direct observation edges, the CFA is traversed from each action α, and for each action β reachable from α, (α, β) is added to D if β observes α. For inter-process observation, we simply iterate over the actions of all other processes and add an indirect observation edge wherever needed. So the data-flow graph can be built in polynomial (quadratic) time in the number of CFA edges.

Example 1. Let us have the simple program from Fig. 1a: its CFA is shown in Fig. 2. The figure also shows two abstract data-flow graphs: the upper one with a precision where some information is tracked about both x and y ($vars(\Pi) = \{x, y\}$); below, we have no information about x ($vars(\Pi) = \{y\}$). Therefore, no edges start from actions assigning x in the second graph. Solid edges are direct observation edges, dashed edges represent inter-process observations.

4.2 Simplifying Statements On-the-Fly Based on Data-Flow

Let Π be the precision of the abstraction, and $G = (A, D, I, \Pi)$ the computed abstract data-flow graph. Let s be a state, $\alpha \in outgoing(s)$: our goal is to decide whether α can be transitively observed later during the program execution in a relevant way. We target reachability properties, so relevant actions are the actions with guard conditions since reachability of error locations of the CFA can only be blocked by conditional statements. We will refer to these relevant actions as *real observers*. Real observers are colored in Fig. 2. Thus, the evaluation of α can be skipped if there is no trace from the current state where a *real observer*

---

[2] On the implementation side, when threads can be created and terminated dynamically, several threads can have the same CFA process. In that case, inter-process observation edges may exist between actions of the same CFA process.

transitively observes α^3. This can be decided using the data-flow graph. To formalize this idea, we introduce the following definitions:

Definition 6. *Let s be an abstract state and p be a process. Let reachable(s, p) denote the set of actions such that $\alpha \in$ reachable(s, p) if there is an abstract trace w in the abstract state space from s with $\alpha \in w$ and $p_\alpha = p$.*

Technically, *reachable*(s, p) is the set of actions that could be executed by p at some point after s. Intuitively, if α is transitively observed by an action β in a trace starting from the current state s, then there is a path in the data-flow graph from α to β only passing through graph nodes (actions) which can still be reached from s by one of the processes. Formally, we define conditions for the enabledness of the data-flow graph edges:

Definition 7. *Let s be an abstract state, and $G = (A, D, I, \Pi)$ an abstract data-flow graph.*

- *An edge $(\alpha_1, \alpha_2) \in D$ is* enabled *in s if $\alpha_1, \alpha_2 \in$ reachable(s, p) for some process p.*
- *An edge $(\alpha_1, \alpha_2) \in I$ is* enabled *in s if $\alpha_1 \in$ reachable(s, p_1) and $\alpha_2 \in$ reachable(s, p_2) for some processes $p_1 \neq p_2$.*

Rephrasing the previous paragraph: if there is a trace from s where α is transitively observed by an action β, then there is a sequence $\alpha_1, ..., \alpha_n$ such that $\alpha_1 = \alpha$, $\alpha_n = \beta$, $(\alpha_i, \alpha_{i+1}) \in D \cup I$ for each $1 \leq i < n$, and (α_i, α_{i+1}) is enabled in s. Using the definition, deciding the enabledness of a data-flow graph edge amounts to answering reachability questions in the state space (see Definition 6) which is also the original purpose of the verification of reachability properties: seemingly, the problem has not become easier. However, *reachable*(s, p) can be over-approximated by checking reachability in the CFA of the program[4].

Example 2. Let us continue our example from Fig. 2 with a precision Π such that $vars(\Pi) = \{x, y\}$ (i.e., the upper data-flow graph in Fig. 2). In the initial state where both processes are in their initial locations (L_0 and L_3), all actions may be reachable in the future by one of the processes since we over-approximate reachability in the state space by reachability in the CFA. Thus, all data-flow graph edges are enabled, so there is a path of enabled data-flow graph edges from both outgoing actions $x = 1$ and $y = x$ to a real observer (e.g., to $[y = 1]$). However, if we have a state where the processes are in locations L_0 and L_4, then $y = x$ can never be reached from this state, so all data-flow graph edges leaving or targeting $y = x$ are disabled. That is, there is no path from $x = 1$ and $x = 0$

---

[3] Note that based on the reflexivity of the transitive observation relation, conditional statements are never simplified.

[4] This over-approximation would be too coarse for the original reachability question of the verification in most cases. However, it can be effectively used for our purposes to answer reachability questions on a lower level.

to a real observer in this state, so these actions are not transitively observed by a real observer, and thus, do not have to be evaluated from this state.

This example also shows a great advantage and novelty of our algorithm over existing cone-of-influence and program slicing techniques: some statements ($x = 1$ in our case) can be removed in certain states even though the same statement may be important and needs to be preserved in other states.

Using an adequate data structure, edge enabledness in the data flow graph can be over-approximated in constant time using CFA reachability information. For indirect edges, CFA reachability information can either be stored in a 2D array (with constant time indexing) or a more memory-efficient, but slightly more over-approximating approach based on strongly connected components can be used (by storing the CFA strongly connected component id number for each CFA edge and comparing these ids on-the-fly). All direct observation edges reachable in G from an action $\alpha \in outgoing(s)$ are enabled based on Definition 7.

For each action $\alpha \in outgoing(s)$, we traverse the data-flow graph from α in the way introduced above. If a real observer is reached, then the value produced by α is used (or at least may be used, c.f., the applied over-approximations), so we evaluate α properly to calculate the successor states $\alpha(s)$. However, if no real observer is reached, then the value is unused, making α unnecessary to evaluate. Instead, the successor state s' can be the state differing from the current state s only in the location of the process of α: $s'(p_\alpha)$ is the target location of α. Specifically, the following method is used to determine the successor states when there is no real observer of α: for the single variable $v \in written(\alpha)$, if $v \in vars(\Pi)$, the original statement assigning a new value to v is replaced by a *havoc v* statement; if $v \notin vars(\Pi)$, the original statement is removed (more precisely replaced with a *no operation* statement that has no effect). Using *havoc* on the variables tracked in the current abstraction is necessary for the refinement step of CEGAR (see Sect. 4.3 for more details).

Algorithm 1 summarizes the presented method of statement simplification based on dynamic data-flow analysis. Line 9 corresponds to the case when simplification is not possible (α has a real observer). Lines 12–13 are applied when α does not have a real observer but the written variable of the action is tracked in the abstraction: so α is replaced by a *havoc* statement. When we have track no information about the written variable in the current abstraction, α is completely eliminated (only the location is updated) in lines 15–17.

Theorem 1 proves that using Algorithm 1 for state space exploration yields correct results, that is, it reaches an error state whenever an error state is reachable with a feasible trace in the original state space. By original state space, we mean the abstract state space explored without the introduced statement simplification (i.e., for each $\alpha \in outgoing(s)$, the successor states $\alpha(s)$ are all explored).

Theorem 1. *Algorithm 1 returns an unsafe verdict whenever an error state is reachable in the concrete state space.*

Algorithm 1: State Space Exploration with Statement Simplification

---

Input: s_0, Π `/* initial state, precision */`
Output: *verdict*   `/* safe/unsafe */`

1   $G \leftarrow$ construct abstract data-flow graph with Π
2   $waitlist \leftarrow \{s_0\}$
3   **while** $waitlist \neq \emptyset$ **do**
4      $s \leftarrow$ remove an item from $waitlist$
5   **if** s *is an error state* **then return** unsafe
6   **else**
7   **foreach** $\alpha \in outgoing(s)$ **do**
8   **if** $\exists$ *path in G of enabled edges in s from α to a real observer* **then**
9              $successors \leftarrow \alpha(s)$
10   **else**
11   **if** $written(\alpha) = \{v\}$ *and* $v \in vars(\Pi)$ **then**
12                 $\alpha' \leftarrow havoc\ v$
13                 $successors \leftarrow \alpha'(s)$
14   **else**
15                 $s' \leftarrow s$
16                 $s'(p_\alpha) \leftarrow$ target location of α
17                 $successors \leftarrow \{s'\}$
18         $waitlist \leftarrow waitlist \cup successors$

19 **return** safe

Proof. A reachable error state in the concrete state space means that the original abstract state space contains a feasible abstract error trace. We prove that if we take *successors* instead of $\alpha(s)$ in a step of the algorithm, then if there is a feasible abstract error trace from s starting with α, there is also a feasible abstract error trace from some $s' \in successors$. We have the following cases:

1. α is transitively observed by a real observer.
   Then α is not simplified, so $successors = \alpha(s)$. Naturally, if there is a feasible abstract error trace from s in the form $\alpha.w$, then w is a feasible abstract error trace from at least one element of $successors = \alpha(s)$.
2. α is not observed transitively by a real observer, and $v \notin vars(\Pi)$ for the single item $v \in written(\alpha)$.[5]
   In this case, α practically has no effect since no information is tracked about v in the current abstraction. So an assignment of v only performs a location update for p_α. This is exactly how s' defined in lines 15–16, so $successors = \alpha(s)$ in this case, as well. Similarly to case 1, there is a feasible abstract error from at least one element of $successors = \alpha(s)$.
3. α is not observed transitively by a real observer, and $v \in vars(\Pi)$: α is replaced by a *havoc* statement.

---

[5] Note that $written(\alpha)$ has exactly one item when α is not transitively observed by a real observer because α must be an assignment then.

A feasible abstract error trace $\alpha.w$ from s implies that there is a concrete state c with $c \models s$ such that $\alpha.w$ is a trace from c to a concrete error state. Note that an unobserved α can be a deterministic or non-deterministic assignment. If we have a non-deterministic assignment, we are back in the previous case since practically, α is not replaced (a *havoc* replaced with a *havoc* on the same variable). So we consider α as a deterministic assignment, that is $\alpha(c) = \{c'\}$. Thus, w is an error trace from c'. Now, if we take α' instead of α, then $c' \in \alpha'(c)$ since a *havoc* means that v can get any value from its domain including the value $c'(v)$ originally assigned by α. Based on the abstraction, for each concrete state $\hat{c} \in \alpha'(c)$ there is an abstract state $\hat{s} \in \alpha'(s)$ such that $\hat{c} \models \hat{s}$. Therefore, for $c' \in \alpha'(c)$, there is an abstract state $s' \in \alpha'(s)$ with $c' \models s'$. This way, w being an error trace from c' implies that w is a feasible abstract error trace from $s' \in successors = \alpha'(s)$.

As the property proven above is preserved in each exploration step, it follows by induction that if a feasible abstract error trace is available from the initial state, then there is a feasible abstract error trace in the state space explored by Algorithm 1, as well, which proves the theorem. $\qquad\square$

4.3 Statement Simplification with CEGAR

Counterexample-Guided Abstraction Refinement (CEGAR) [22] is an abstraction-based model checking algorithm, starting from a coarse abstraction and iteratively refining it until it can prove or disprove the analyzed property. Its core is the *CEGAR-loop* consisting of the *abstractor* and the *refiner*. The abstractor builds the abstract state space over an abstract domain with a given precision. Since this is an over-approximation of the original concrete state space, if no abstract error state is reachable, the concrete model is also safe. On the other hand, when an abstract error is reachable, the refiner checks whether it is a *feasible* or a *spurious* abstract counterexample. The counterexample is an alternating sequence of abstract states and actions from the initial abstract state to an abstract error state. The refiner checks whether this trace is feasible or not, that is, whether there is a concrete variable assignment for each state of the trace that does not contradict the abstract state expressions and the actions of the trace. If the abstract trace is feasible, the program is unsafe. If the trace is spurious, the precision is refined (i.e., the refiner provides a new refined precision that could better establish the safety or reveal the faulty behavior of the program). The abstract state space is built with this refined precision in the next iteration.

Our proposed algorithm can be used by the abstractor of CEGAR for abstract state space exploration. However, it is important that a potential counterexample provided to the refiner must contain the original actions even if our algorithm simplified them during the state space exploration. For a reason, consider a program with a single process which is *Process* p_1 from Fig. 2, but let the action from L_1 to L_2 be $y = x$. Let our precision only track information about x: $vars(\Pi) = \{x\}$. An error state is reachable in the abstract state space with the clearly spurious trace ($x = 1$, $y = x$, $[y \neq 1]$). However, our algorithm simplifies $x = 1$ and $y = x$ since they are not observed by a conditional action with this precision. If the refiner only sees the simplified actions in the trace, i.e., (*havoc x*, *no operation*, $[y \neq 1]$), the contradiction cannot be spotted. Concluding that the counterexample is feasible, it would give a wrong *unsafe* verdict (instead of spotting the contradiction and providing a better refined precision).

Our algorithm presented previously can also be used in verification algorithms other than CEGAR. In such a case, it may be possible to drop lines 11–14 of Algorithm 1 and always use lines 15–17 to define the successor state when α is not observed transitively by a real observer. However, we focus on CEGAR as the base algorithm in this paper, making the *havoc* necessary when the assignment of a variable in the precision is simplified. If the successor state is defined as in lines 15–16 instead of using a *havoc*, the refiner may see a contradiction. For example, assume that the value of variable x is explicitly tracked in a CEGAR iteration, and the abstractor finds a counterexample trace. The trace contains a state s where $x = 0$, and the next action α assigns 1 to x. However, our algorithm noticed that α cannot be transitively observed by any real observer, so it skipped the evaluation of the statement of α. Based on lines 15–16 of the algorithm, the value of x would be the same (namely 0) in the state s' after α in the trace. Then, the refiner finds a contradiction here as the value of x cannot be 0 after an action that assigns 1 to x (we have seen in the previous paragraph that the counterexample must contain the original actions). On the other hand, the precision could not be refined based on this misleading contradiction, and the CEGAR algorithm could get stuck in endless iterations.

Applying a *havoc* statement instead of the unevaluated assignment expression overcomes this problem, as the *havoc* statement covers the behavior of the original assignment whatever value it would assign. Going back to our example, the *havoc* statement would erase the value of x from s', so it is not a contradicting state after α. Evaluating the *havoc* statement is still a simple task, so it is still worth replacing the original assignments with it.

It is worth mentioning that our algorithm cannot introduce new spurious counterexamples and degrade performance this way. Intuitively, guard conditions cannot get enabled as a side-effect of our algorithm since Algorithm 1 only simplifies statements that are not observed transitively by any conditional statement. Thus, the evaluation of guard conditions is not affected, so new (spurious) counterexamples cannot emerge. As for originally feasible counterexamples, they remain feasible with our algorithm as feasible traces are always available in the abstract state space explored by our algorithm based on the proof of Theorem 1.

5   Experimental Evaluation

In this section, we evaluate the efficiency of our algorithmic contributions. The goal of our experiment is to evaluate the performance of our proposed dynamic statement simplification algorithm. We refer to our novel algorithm as a dynamic cone-of-influence-based statement simplification algorithm (or DCOI for short) in our experiments. We compare our algorithm to a baseline using static cone-of-influence (SCOI) to see how much further reduction is achieved (that is when DCOI and SCOI are both enabled). Since DCOI can do the job of SCOI so to say (and thus, completely replace SCOI), it is also meaningful to investigate the performance with only DCOI while SCOI is disabled. We are interested in different abstract domains, therefore we investigate the effect of our proposed algorithm in two abstract domains frequently used in state-of-the-art verification tools [11]: explicit-value abstraction (later EXPL) [17] and (Cartesian) predicate abstraction (later PRED) [9].

We implemented our algorithm as an open-source extension of the THETA verification tool [48] which already had a built-in CEGAR algorithm and a static cone-of-influence preprocessing step, and had prior support for multi-threaded C programs including a partial order reduction algorithm [8]. We also compare our results to other state-of-the-art verifiers using abstract state space exploration.

In our experiments, we use a *static* partial order reduction (POR) algorithm [2,8] during the state space exploration in a way that POR is applied first as a filter on the outgoing actions of states, and then the DCOI technique is used to simplify the subset of outgoing actions selected by POR. Since the applied static POR only takes actions into account that might be executed later, the soundness of the POR algorithm is not affected by our proposed algorithm (as the effect of DCOI is materialized in the already explored part of the state space).

Research Questions. To evaluate the presented algorithm, we aim to answer the following research questions concerning metrics relevant to it:

RQ1 What proportion of statements can be simplified or completely eliminated using the proposed algorithm?
RQ2 How is the time of successor state calculation affected by our algorithm?
RQ3 How is the overall verification performance affected by the algorithm?
RQ4 What practical performance improvement can we observe on programs where theoretically exponential gain is expected?

Experimental Configuration. In our experiments, we executed different configurations of THETA over a set of input programs written in C from the concurrency safety reachability category benchmark suite[6] of SV-COMP [11] (715 tasks) that is parsable by THETA (598) for RQ1-RQ3 and a direct implementation of Fig. 1b for RQ4 with $N = 2^{0 \leq i \leq 7}$. We executed 6 configurations on the SV-COMP benchmarks: both abstract domains (EXPL, PRED) with the three different cone-of-influence methods (SCOI, SCOI+DCOI, DCOI). The benchmark tests were executed on virtual machines with Intel Core (Haswell) processors, 2 dedicated CPU cores were allocated to each task. Each verification task had a time limit of 900 s (1800 s for RQ4) and a memory limit of 15 GB. We used

---

[6] https://gitlab.com/sosy-lab/benchmarking/sv-benchmarks.

a sequence interpolation-based refinement strategy for the refinement step of CEGAR, and depth-first state space exploration with thread-safe large-block encoding and a static abstraction-based partial order reduction algorithm [8] in the abstraction phase. We used atoms as the basis of predicate splitting for the predicate domain; and we used a maximum number of enumerated successor states (maxenum) of 1 for the explicit domain [32]. Our backend SMT solver was Z3. For the exponential gain program, we used the predicate abstract domain with an initial precision obtained by extracting branching conditions from the program; and we also applied a simple static partial order reduction algorithm [2] after applying DCOI (the same POR algorithm is used when DCOI is disabled for a fair comparison).

5.1 Experiment Results

In the concurrency safety benchmark suite, THETA was able to parse 602 programs. No configuration provided incorrect results (where the verdict reported by THETA differs from the expected result). Table 1 shows the results for different metrics aggregated by configuration. For a fair comparison, the aggregated values are calculated over the common subset of correctly solved tasks by abstract domain: a common subset of 332 tasks was solved with the configurations using explicit-value abstraction, and 350 with predicate abstraction.

The *simplified by DCOI* column shows the average proportion of simplified statements (including statements replaced by *havoc* and completely eliminated statements) simplified by DCOI compared to all statements. The *successor calculation* and *CPU time* columns are the sum of successor state calculation and CPU times of commonly solved tasks.

The results confirm the reduction potential of our algorithm: configurations using DCOI greatly outperform (depending on the abstract domain) the baselines without DCOI in terms of both successor state calculation and overall verification performance. It is also in line with our expectations that using DCOI without SCOI leads to slightly better performance since DCOI can also eliminate the statements removed by SCOI with a minor overhead while the time of SCOI is completely spared. Let us interpret the results by answering the research questions:

RQ1 DCOI simplifies 19.6% of all statements on average (14% completely eliminated, while 5.6% replaced by *havoc* with explicit-value abstraction; 14.4% and 5.2% respectively with predicate abstraction). This confirms the relevance of our method: a significant subset of statements is unnecessary in certain thread interleavings for verifying the given property of the program.

RQ2 Our algorithm greatly reduces the time of successor state calculation as testified by the results in Table 1: by 29.9% with explicit abstraction and 31.8% with predicate abstraction. A significant part of successor state calculation is taken by SMT-solvers solving SMT problems (especially when using predicate abstraction). Thus, the overall system load is significantly decreased by reducing the SMT problem solving time.

RQ3 Overall performance is also improved (see CPU time in Table 1), especially for predicate abstraction: DCOI reduces the overall CPU time compared to

Table 1. Different metrics of the evaluation

domain	coi	simplified by DCOI	successor calculation	CPU time	solved tasks
	SCOI	0%	1254 s	5581 s	332
EXPL	SCOI+DCOI	19.5%	880 s	5548 s	332
	DCOI	19.6%	879 s	5376 s	334
	SCOI	0%	2168 s	0496 s	352
PRED	SCOI+DCOI	19.7%	5289 s	5102 s	358
	DCOI	19.6%	5118 s	4582 s	358

the baseline by 3.7% using explicit abstraction, and by 14.6% using predicate abstraction. It was our expectation to have better improvement with predicate abstraction since it is more costly to compute which tracked predicates (or their negations) are entailed by the previous abstract state and the current action. Thus, successor state calculation takes a greater portion of the verification in predicate abstraction leading to a greater impact of our algorithm. The number of solved tasks is only slightly increased (i.e., the number of tasks where the baseline configuration hit the timeout, but our approach enabled THETA to verify the program within the time limit) probably because the complexity of input tasks is not linearly increasing. The overhead of our algorithm (the time of building and traversing the data-flow graph) is not huge though not completely negligible: 197 s and 206 s for SCOI+DCOI and DCOI, respectively, aggregated for all tasks with explicit-value analysis which is 3.6% and 3.8% of all CPU time. Similarly, our algorithm ran for a total of 342 and 345 s with predicate abstraction taking 1% of all CPU time in both cases.

RQ4 Even though the baseline used the same static COI and the same partial order reduction algorithm, it could only solve the three smallest tasks in the set (up to $N = 4$) within the time limit, whereas DCOI was able to verify 8 tasks, up to $N = 128$ as seen in Fig. 3. Indeed, our algorithm scales very well in some cases. We tried to solve these tasks with state-of-the-art verifiers, as well. Even the best tools in SV-COMP 2024[7] (DARTAGNAN and ULTIMATE GEMCUTTER, winner and second place in the concurrency category of SV-COMP'24 [11]) cannot solve these tasks for $N > 4$ within the time limit which highlights the potential of our algorithm.

Comparison with the State-of-the-Art. We also compare the average performance of our solution to other state space exploration algorithms in state-of-the-art verifiers. We select the best performing verifiers from SV-COMP 2024 concurrency category [11] that use some kind of abstract state space exploration algorithm. Two such state-of-the-art tools are CPACHECKER [7] and PICHECKER [47]. Other successful tools in SV-COMP are either bounded model checkers (such as DARTAGNAN [41], DEAGLE [34], and CSEQ [25]) that would be unfair choices

---

[7] The official results of SV-COMP 2025 are not published at the time of writing.

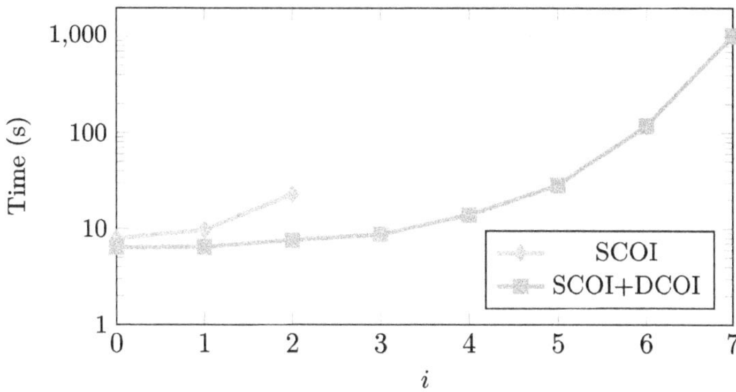

Fig. 3. Execution time given i for $N := 2^i$ in Fig. 1b

for comparison with a complete model checking algorithm; or use a conceptually different trace abstraction algorithm (such as ULTIMATE AUTOMIZER [35], GEMCUTTER [38], and TAIPAN [26]); or use some advanced algorithm or tool selection strategies without implementing own analyses (such as PESCO [46], and GRAVES [40]).

We executed the two verifiers on the same SV-COMP tasks, on the same hardware with the same limits. CPACHECKER uses a standard state space exploration technique for multi-threaded programs combined with a BDD analysis [13] and achieved to solve 346 tasks correctly[8]. PICHECKER is built on CPACHECKER and has multiple analyses for concurrent software [47]. One that uses CEGAR and Craig interpolation could verify 246 tasks while its main method, a BDD analysis with an elaborate partial order reduction algorithm could verify 388 tasks. Our solution solving 358 tasks thus ranks second among these similar analyses. While our contribution does not bring THETA to the first place among these tools, it reduces the advantage of PICHECKER. As a reference, the most solved tasks by a single tool was 452 in the SV-COMP 2024 concurrency reachability category [11], though that tool used bounded model checking.

5.2   Threats to Validity

Internal Validity. We used BenchExec [19] to ensure the accuracy of our experiments executed on virtual machines in our university's cloud computing platform. External factors such as shared resources may have influenced the results.

External Validity. The SV-COMP benchmark suite is considered a de facto standard for academic benchmarking in software verification. THETA can only parse a limited subset of SV-COMP concurrent benchmark programs which further reduces generalizability. However, there might be more redundant model ele-

---

[8] CPACHECKER also has a predicate analysis for concurrent software [16] but the algorithm is a bit dated and this analysis can only verify 159 tasks.

ments in real-world software than in the simplified programs of the SV-COMP benchmarks, making our technique disadvantaged on the benchmark set.

Construct Validity. Evaluation metrics were carefully chosen to accurately describe the performance of our algorithm: both *end-user* statistics (such as CPU time, number of solved tasks) and *backend-related* information (such as the ratio of simplified statements, successor state calculation time) were used.

6 Conclusion

In this paper, we have presented a novel statement reduction algorithm based on dynamic data-flow analysis to aid abstract state space exploration of concurrent programs. Our method is based on a similar idea to cone-of-influence algorithms, however, our algorithm performs a more fine-grained analysis resulting in more extensive reduction of model elements. We have proven its correctness and discussed its integration into the abstraction-based verification algorithm CEGAR. The evaluation of the algorithm shows that our approach can simplify or completely eliminate a great proportion of statements which leads to a significant improvement in both successor state calculation time and overall verification time, especially in cases where successor state calculation takes a significant proportion of verification time, such as in the case of predicate abstraction. Therefore, the presented algorithm is worth implementing in a model checking tool that verifies concurrent software.

Funding Information. This research was partially funded by the 2024-2.1.1-EKÖP-2024-00003 University Research Scholarship Programme under project numbers EKÖP-24-2-BME-118 and EKÖP-24-3-BME-{159,213}, and the Doctoral Excellence Fellowship Programme under project numbers 400434/2023 and 400443/2023; with the support provided by the Ministry of Culture and Innovation of Hungary from the NRDI Fund.

References

1. Abdulla, P.A., Aronis, S., Jonsson, B., Sagonas, K.: Optimal dynamic partial order reduction, pp. 373–384. ACM (2014). https://doi.org/10.1145/2535838.2535845
2. Abdulla, P., Aronis, S., Jonsson, B., Sagonas, K.: Comparing source sets and persistent sets for partial order reduction. In: Aceto, L., Bacci, G., Bacci, G., Ingólfsdóttir, A., Legay, A., Mardare, R. (eds.) Models, Algorithms, Logics and Tools. LNCS, vol. 10460, pp. 516–536. Springer, Cham (2017). https://doi.org/10.1007/978-3-319-63121-9_26
3. Abdulla, P.A., Aronis, S., Jonsson, B., Sagonas, K.: Source sets: a foundation for optimal dynamic partial order reduction. J. ACM **64**(4), 25:1–25:49 (2017). https://doi.org/10.1145/3073408
4. Agarwal, P., Chatterjee, K., Pathak, S., Pavlogiannis, A., Toman, V.: Stateless model checking under a reads-value-from equivalence. Lecture Notes in Computer Science, vol. 12759, pp. 341–366. Springer (2021). https://doi.org/10.1007/978-3-030-81685-8_16

5. Agrawal, H., Horgan, J.R.: Dynamic program slicing. pp. 246–256. ACM (1990). https://doi.org/10.1145/93542.93576

6. Aronis, S., Jonsson, B., Lång, M., Sagonas, K.: Optimal dynamic partial order reduction with observers. In: Beyer, D., Huisman, M. (eds.) TACAS 2018. LNCS, vol. 10806, pp. 229–248. Springer, Cham (2018). https://doi.org/10.1007/978-3-319-89963-3_14

7. Baier, D., et al.: Cpachecker 2.3 with strategy selection - (competition contribution). LNCS, vol. 14572, pp. 359–364. Springer (2024). https://doi.org/10.1007/978-3-031-57256-2_21

8. Bajczi, L., et al.: Theta: Abstraction based techniques for verifying concurrency (competition contribution). LNCS, vol. 14572, pp. 412–417. Springer, Cham (2024). https://doi.org/10.1007/978-3-031-57256-2_30

9. Ball, T., Podelski, A., Rajamani, S.K.: Boolean and cartesian abstraction for model checking C programs. In: Margaria, T., Yi, W. (eds.) TACAS 2001. LNCS, vol. 2031, pp. 268–283. Springer, Heidelberg (2001). https://doi.org/10.1007/3-540-45319-9_19

10. Berezin, S., Campos, S., Clarke, E.M.: Compositional reasoning in model checking. In: de Roever, W.-P., Langmaack, H., Pnueli, A. (eds.) COMPOS 1997. LNCS, vol. 1536, pp. 81–102. Springer, Heidelberg (1998). https://doi.org/10.1007/3-540-49213-5_4

11. Beyer, D.: State of the art in software verification and witness validation: SV-COMP 2024. LNCS, vol. 14572, pp. 299–329. Springer (2024). https://doi.org/10.1007/978-3-031-57256-2_15

12. Beyer, D., Dangl, M., Wendler, P.: A unifying view on SMT-based software verification. J. Autom. Reason. 60(3), 299–335 (2017). https://doi.org/10.1007/s10817-017-9432-6

13. Beyer, D., Friedberger, K.: A light-weight approach for verifying multi-threaded programs with CPAchecker. EPTCS 233, 61–71 (2016). https://doi.org/10.4204/EPTCS.233.6

14. Beyer, D., Henzinger, T.A., Théoduloz, G.: Configurable software verification: concretizing the convergence of model checking and program analysis. In: Damm, W., Hermanns, H. (eds.) CAV 2007. LNCS, vol. 4590, pp. 504–518. Springer, Heidelberg (2007). https://doi.org/10.1007/978-3-540-73368-3_51

15. Beyer, D., Keremoglu, M.E.: CPACHECKER: a tool for configurable software verification. In: Gopalakrishnan, G., Qadeer, S. (eds.) CAV 2011. LNCS, vol. 6806, pp. 184–190. Springer, Heidelberg (2011). https://doi.org/10.1007/978-3-642-22110-1_16

16. Beyer, D., Keremoglu, M.E., Wendler, P.: Predicate abstraction with adjustable-block encoding, pp. 189–197. IEEE (2010)

17. Beyer, D., Löwe, S.: Explicit-Value Analysis Based on CEGAR and Interpolation. CoRR abs/1212.6542 (2012)

18. Beyer, D., Löwe, S., Wendler, P.: Sliced path prefixes: an effective method to enable refinement selection. LNCS, vol. 9039, pp. 228–243. Springer (2015). https://doi.org/10.1007/978-3-319-19195-9_15

19. Beyer, D., Löwe, S., Wendler, P.: Reliable benchmarking: requirements and solutions. Int. J. Softw. Tools Technol. Transfer 21(1), 1–29 (2017). https://doi.org/10.1007/s10009-017-0469-y

20. Blanc, N., Kroening, D.: Race analysis for systemc using model checking. ACM Trans. Design Autom. Electr. Syst. 15(3), 21:1–21:32 (2010). https://doi.org/10.1145/1754405.1754406

21. Bozzano, M., Cimatti, A., Tapparo, F.: Symbolic fault tree analysis for reactive systems. In: Namjoshi, K.S., Yoneda, T., Higashino, T., Okamura, Y. (eds.) ATVA 2007. LNCS, vol. 4762, pp. 162–176. Springer, Heidelberg (2007). https://doi.org/10.1007/978-3-540-75596-8_13

22. Clarke, E.M., Grumberg, O., Jha, S., Lu, Y., Veith, H.: Counterexample-guided abstraction refinement for symbolic model checking. J. ACM 50(5), 752–794 (2003). https://doi.org/10.1145/876638.876643

23. Clarke, E.M., Grumberg, O., Long, D.E.: Model checking and abstraction. ACM Trans. Program. Lang. Syst. 16(5), 1512–1542 (1994). https://doi.org/10.1145/186025.186051

24. Clarke, E.M., Klieber, W., Nováček, M., Zuliani, P.: Model checking and the state explosion problem. In: Meyer, B., Nordio, M. (eds.) LASER 2011. LNCS, vol. 7682, pp. 1–30. Springer, Heidelberg (2012). https://doi.org/10.1007/978-3-642-35746-6_1

25. Coto, A., Inverso, O., Sales, E., Tuosto, E.: A prototype for data race detection in CSeq 3. In: TACAS 2022. LNCS, vol. 13244, pp. 413–417. Springer, Cham (2022). https://doi.org/10.1007/978-3-030-99527-0_23

26. Dietsch, D., Heizmann, M., Klumpp, D., Schüssele, F., Podelski, A.: Ultimate taipan and race detection in ultimate - (competition contribution). LNCS, vol. 13994, pp. 582–587. Springer (2023). https://doi.org/10.1007/978-3-031-30820-8_40

27. Dietsch, D., Heizmann, M., Musa, B., Nutz, A., Podelski, A.: Craig vs. newton in software model checking, pp. 487–497. ACM (2017). https://doi.org/10.1145/3106237.3106307

28. Dwyer, M.B., Clarke, L.A.: Data Flow Analysis for Verifying Properties of Concurrent Programs, pp. 62–75. ACM (1994). https://doi.org/10.1145/193173.195295

29. Eilers, M., Dardinier, T., Müller, P.: Commcsl: Proving information flow security for concurrent programs using abstract commutativity. Proc. ACM Program. Lang. 7(PLDI), 1682–1707 (2023). https://doi.org/10.1145/3591289

30. Farzan, A., Klumpp, D., Podelski, A.: Stratified commutativity in verification algorithms for concurrent programs. Proc. ACM Program. Lang. 7(POPL), 1426–1453 (2023). https://doi.org/10.1145/3571242

31. Flanagan, C., Qadeer, S.: Predicate abstraction for software verification, pp. 191–202. ACM (2002). https://doi.org/10.1145/503272.503291

32. Hajdu, Á., Micskei, Z.: Efficient strategies for CEGAR-based model checking. J. Autom. Reason. 64(6), 1051–1091 (2019). https://doi.org/10.1007/s10817-019-09535-x

33. Harman, M., Hierons, R.M.: An overview of program slicing. Softw. Focus 2(3), 85–92 (2001). https://doi.org/10.1002/swf.41

34. He, F., Sun, Z., Fan, H.: Deagle: an smt-based verifier for multi-threaded programs (competition contribution). In: International Conference on Tools and Algorithms for the Construction and Analysis of Systems, pp. 424–428. Springer (2022)

35. Heizmann, M., et al.: Ultimate automizer and the commuhash normal form - (competition contribution). LNCS, vol. 13994, pp. 577–581. Springer (2023). https://doi.org/10.1007/978-3-031-30820-8_39

36. Henzinger, T.A., Jhala, R., Majumdar, R., Sutre, G.: Software verification with BLAST. In: Ball, T., Rajamani, S.K. (eds.) SPIN 2003. LNCS, vol. 2648, pp. 235–239. Springer, Heidelberg (2003). https://doi.org/10.1007/3-540-44829-2_17

37. Huang, J.: Stateless model checking concurrent programs with maximal causality reduction. pp. 165–174. ACM (2015). https://doi.org/10.1145/2737924.2737975

38. Klumpp, D., Dietsch, D., Heizmann, M., Schüssele, F., Ebbinghaus, M., Farzan, A., Podelski, A.: ULTIMATE GEMCUTTER and the axes of generalization. In: TACAS 2022. LNCS, vol. 13244, pp. 479–483. Springer, Cham (2022). https://doi.org/10.1007/978-3-030-99527-0_35
39. Korel, B., Rilling, J.: Dynamic program slicing methods. Inf. Softw. Technol. **40**(11–12), 647–659 (1998)
40. Leeson, W., Dwyer, M.B.: Graves-cpa: a graph-attention verifier selector (competition contribution). In: International Conference on Tools and Algorithms for the Construction and Analysis of Systems, pp. 440–445. Springer (2022)
41. Ponce-de-León, H., Haas, T., Meyer, R.: DARTAGNAN: leveraging compiler optimizations and the price of precision (competition contribution). In: TACAS 2021. LNCS, vol. 12652, pp. 428–432. Springer, Cham (2021). https://doi.org/10.1007/978-3-030-72013-1_26
42. Loiacono, C., et al.: Fast cone-of-influence computation and estimation in problems with multiple properties. pp. 803–806. EDA Consortium San Jose, CA, USA/ACM DL (2013). https://doi.org/10.7873/DATE.2013.170
43. Nanda, M.G., Ramesh, S.: Slicing concurrent programs, pp. 180–190. ACM (2000). https://doi.org/10.1145/347324.349121
44. Peled, D.: Ten years of partial order reduction. In: Hu, A.J., Vardi, M.Y. (eds.) CAV 1998. LNCS, vol. 1427, pp. 17–28. Springer, Heidelberg (1998). https://doi.org/10.1007/BFb0028727
45. Peled, D.A., Valmari, A., Kokkarinen, I.: Relaxed visibility enhances partial order reduction. Formal Methods Syst. Des. **19**(3), 275–289 (2001). https://doi.org/10.1023/A:1011202615884
46. Richter, C., Hüllermeier, E., Jakobs, M., Wehrheim, H.: Algorithm selection for software validation based on graph kernels. Autom. Softw. Eng. **27**(1), 153–186 (2020). https://doi.org/10.1007/S10515-020-00270-X
47. Su, J., Yang, Z., Xing, H., Yang, J., Tian, C., Duan, Z.: Pichecker: A POR and interpolation based verifier for concurrent programs (competition contribution). LNCS, vol. 13994, pp. 571–576. Springer (2023). https://doi.org/10.1007/978-3-031-30820-8_38
48. Tóth, T., Hajdu, A., Vörös, A., Micskei, Z., Majzik, I.: THETA: a Framework for Abstraction Refinement-Based Model Checking, pp. 176–179 (2017). https://doi.org/10.23919/FMCAD.2017.8102257

Author Index

GPSR Compliance
The European Union's (EU) General Product Safety Regulation (GPSR) is a set
of rules that requires consumer products to be safe and our obligations to
ensure this.

If you have any concerns about our products, you can contact us on

ProductSafety@springernature.com

In case Publisher is established outside the EU, the EU authorized
representative is:

Springer Nature Customer Service Center GmbH
Europaplatz 3
69115 Heidelberg, Germany